MW00452763

A Primer for Sampling Solids, Liquids, and Gases

ASA-SIAM Series on Statistics and Applied Probability

The ASA-SIAM Series on Statistics and Applied Probability is published jointly by the American Statistical Association and the Society for Industrial and Applied Mathematics. The series consists of a broad spectrum of books on topics in statistics and applied probability. The purpose of the series is to provide inexpensive, quality publications of interest to the intersecting membership of the two societies.

Smith, P. L., *A Primer for Sampling Solids, Liquids, and Gases: Based on the Seven Sampling Errors of Pierre Gy*

Meyer, M. A. and Booker, J. M., *Eliciting and Analyzing Expert Judgment: A Practical Guide*

Latouche, G. and Ramaswami, V., *Introduction to Matrix Analytic Methods in Stochastic Modeling*

Peck, R., Haugh, L., and Goodman, A., *Statistical Case Studies: A Collaboration Between Academe and Industry, Student Edition*

Peck, R., Haugh, L., and Goodman, A., *Statistical Case Studies: A Collaboration Between Academe and Industry*

Barlow, R., *Engineering Reliability*

Czitrom, V. and Spagon, P. D., *Statistical Case Studies for Industrial Process Improvement*

A Primer for Sampling Solids, Liquids, and Gases

Based on the Seven Sampling Errors of Pierre Gy

Patricia L. Smith
Alpha Stat Consulting Company
Lubbock, Texas

ASA

Society for Industrial and Applied Mathematics
Philadelphia, Pennsylvania

American Statistical Association
Alexandria, Virginia

10 9 8 7 6 5 4 3 2 1

Library of Congress Cataloging-in-Publication Data

Smith, Patricia L.
A primer for sampling solids, liquids, and gases : based on the seven sampling errors of Pierre Gy / Patricia L. Smith.
p. cm. – (ASA-SIAM series on statistics and applied probability)
Includes bibliographical references and index.
ISBN 0-89871-473-7
1. Bulk solids–Sampling. 2. Sampling. 3. Gases–Analysis. 4. Liquids–Analysis. I. Title. II. Series.

TA418.78 .S65 2000
620'.001'51952–dc21

00-061250

The photograph of Patricia L. Smith appears courtesy of the photographer, Amy Ashmore.

To Phil
for support and encouragement

Contents

Preface

In a 1962 paper in *Technometrics* (p. 330), Duncan states, "Randomness in the selection of original increments is indeed the number one problem in the sampling of bulk material." In my work as an industrial statistician, I certainly found this to be true. In fact, I was often at a loss in many practical sampling situations. I knew I needed help, and fortunately, I got it.

I was first introduced to the sampling theory of Pierre Gy by Francis Pitard, a student and practitioner of Gy's work. I was at once excited and overwhelmed. I was excited because I was exposed to a new and structured approach to sampling that would help me solve the practical problems I encountered in my work. At the same time, I was overwhelmed by the theory, the terminology, and the scope. The theory was totally different from, but not in conflict with, my formal statistical training. My graduate education in sampling theory addressed only situations in which the sampling units were well defined (people, manufactured parts, etc.). I did not learn in these statistics courses how to sample bulk materials (soils from the ground or in piles, powders, liquids, etc.), nor how much to sample. *Gy's theory, however, provided me with*

1. *a* ***structured approach*** *from which I could break down a sampling problem into component parts and*

2. ***basic principles*** *that I could apply to any sampling situation.*

Every subject area uses words in a specialized way, and sometimes the language is a stumbling block. To minimize possible confusion, I have used synonyms for some of Gy's words. I hope to convey his intuitive meaning while at the same time preserving his original ideas. I do not wish to trivialize or even oversimplify this complicated area, but I want to present the ideas in lay terms so that the *concepts and principles* are easily grasped and therefore more likely to be *applied*. Consequently, I also sacrifice technical correctness at times to convey the essentials.

Pierre Gy has written a number of papers and consolidated his work in several books. For simplicity, I quote mainly from his most recent books, one in 1992, consolidating his theory in its entirety, and his latest one in 1998, an abbreviated version that highlights the major elements. Francis Pitard has presented the essence of Gy's theory in a recent book and examined the idea of chronostatistics, applying part of Gy's theory to control charting. With his publications in the area of spatial statistics, Noel Cressie has "legitimized," from a statistical viewpoint, the variogram, one of Gy's key tools for studying process variation. All of these works are sophisticated and require time and effort to understand and apply. In addition to these books and papers in technical journals, there is a wealth of articles in trade journals addressing specific sampling situations. Advertisements for sampling equipment abound. A search on the Internet can locate sampling information, documents, and vendors.

For the sampling practitioner, I saw the need for a short and useable handbook that would concentrate on the "how to's" and rely on intuitive explanations for the "why's." If intuitive explanations fall short, the inquisitive reader should go to Gy (1992, 1998) or Pitard (1993) to fill in the details. *This book is an introduction only.* It is designed for people in the field to serve as a guide that presents Gy's principles for use in day-to-day applications. It is also designed to serve as an overview and introduction to Gy's theory from which a more detailed study can be launched. The focus is on representative sampling in industrial, process, and lab settings. I do not address special situations, such as environmental contamination, which may require judgmental sampling or location sample selection based on existing data. However, some of the principles in Chapter 3 can be used to help characterize spatial distributions. While many examples reflect my industrial background, Gy's principles apply to any sampling situation in which representative sampling is desired.

Footnotes serve two purposes. In some cases, they provide theoretical details. These might be of interest to readers who want more technical information. Other footnotes provide expanded explanations of words or phrases to clarify the text. These should be useful to all readers. Appendix A provides a brief overview of the material and can actually be read first to get an idea of the type of thinking involved in this approach to sampling. Chapter 1 has a brief introduction, which provides background and motivation for the rest of the book. Chapters 2–4 give detail, and each has a summary section that can be used as a quick reference once the ideas are understood. I like Chapter 3 the best because it discusses the practical details of physical sampling and the difficulty of getting random samples from solid, liquid, or gas lots. The material in this chapter is what was most fascinating to me in my initial exposure to this subject. It is also the easiest to understand and apply. Chapter 5 suggests an overall strategy for implementing the ideas presented. Appendices B and D expand the technical detail from earlier chapters and are fairly specialized. Appendix C gives a practical way to obtain a random sample from a sequential set of items. Appendix E describes some simple experiments that can be performed with commonplace materials. I strongly encourage readers to try these because they illustrate in a straightforward way the difficulties of sampling bulk materials and certain liquids. While trying to reduce sampling errors, theoretical principles must sometimes be compromised in practice to ensure the safety of the person taking the samples. Thus, I include some information on safety in Appendix F. While the environmental impact of sampling should always be considered, I do not discuss this specialized subject.

Several people reviewed drafts in various stages of development. I am especially grateful to Francis Pitard. From his carefully crafted remarks and the detailed discussions I had with him, I was able to clarify ambiguous ideas and present them in a more complete framework. His support and encouragement throughout are very much appreciated. Lynn Charney and Greg Stenback provided valuable comments. Lynn furnished an overall perspective on organization and readability. Greg examined details including definitions and formulas. He made suggestions for rewording to clarify intended meanings and help avoid misinterpretations. I even found a way to put him in the references. Tom Welker contributed his expertise on liquid and gas sampling. I would also like to thank the reviewers of the final manuscript for their helpful comments, enabling further refinements and improvements.

Patricia L. Smith

Foreword

Today, the entire statistics world is acquainted with Dr. Pierre Gy's gem called the "sampling theory," which is the only sampling theory that is complete and founded on solid, scientific ground. After teaching several hundred courses around the world to many thousands of people on the virtues of the sampling theory, I came to the conclusion that three kinds of professionals attend these courses.

1. Some are sent to the course by their bosses, and what they learn helps them later on to perform their jobs better.

2. A few, usually managers, come voluntarily and what they learn opens their eyes to causes of problems, often related to sampling issues, and the awesome invisible cost they generate.

3. On rare occasions, a scientist with solid mathematical and statistical background will attend, and what he or she learns opens new doors to this remarkable way of thinking. Dr. Patricia L. Smith is one of these.

Pat attended several different versions of the course about 10 years ago. Then, within Shell, she presented lectures on her own. Soon, she was capable of presenting the entire course by herself. Finally, one day, she made a comment that annoyed me at first: "There is a need for a simple introductory book, using a language most people can understand." For a long time, I was reluctant to accept this idea, arguing that many papers published around the world had enormously damaged the original work by oversimplifying a complex subject they did not understand, in depth, before pretending to bring beneficial contribution. Furthermore, others were quick to criticize that work because so much was obviously missing in those naïve articles.

My fear was that a simplistic book may wrongly give the impression that the sampling theory can be simplified to the point where anyone can have access to it without making the effort to acquire enough maturity on the subject.

Nevertheless, Pat wrote her book, and I reviewed it several times. Along the way, my original fear vanished. Pat's book had the merit to expose the complex concepts of the sampling theory with an eye trained in a totally different way from the one trained in the mining industry. Her book can be used as a quick introduction to the necessary concepts in the chemical industry, the oil industry, the food industry, and in environmental science. Her success is mainly due to the pragmatic nature of the many examples she has selected. They are real world, and many people will like this. Therefore, I hope her book will help many to grasp the importance of understanding concepts first, before going into the wonderful depth of the original work. The sampling theory is universal, and if well presented, it can help anyone, in any industry, to solve complex sampling problems.

Finally, Pat never gives up. Over time, she can travel in the most intricate subtleties of the sampling theory, then ask most of the relevant questions, and perhaps add new ideas. For what the sampling theory really is, Dr. Patricia L. Smith is the kind of contributor we need to encourage and support. Pat, you are welcome to the club of those who strongly believe in the extraordinary value of the work of Dr. Pierre Gy.

Francis F. Pitard
President of Francis Pitard Sampling Consultants, L.L.C.

List of Terms and Symbols

Analytical error	Measurement error in the analysis of chemical or physical properties.
Bias	The difference between the average amounts from many samples and the "true" lot value.
c	Content (weight proportion) of the component of interest.
Coliwasa	COmposite LIquid WAste SAmpler: a tool used for getting a vertical core from a liquid lot.
Composite sample	A sample obtained by combining several distinct subsamples or increments.
Constitution heterogeneity (CH)	A difference in the amount of the component of interest between individual fragments, particles, or molecules.
Control chart	A time plot with upper and lower control limits determined by the natural variation in the process.
Delimitation error (DE)	The error incurred when the boundary of the sample to be taken is not defined according to the principle of correct sampling.
Distribution heterogeneity (DH)	A difference in the amount of the component of interest between groups of fragments, particles, or molecules.
Extraction error (EE)	The error incurred when the sample taken is not the one that was defined.
Fundamental error (FE)	The difference between the amount of the constituent of interest in the sample and what the "true" value is in the lot.

Grab sample	A sample obtained conveniently instead of by following the principle of correct sampling.
Grouping and segregation error (GSE)	The error arising from not sampling the units in the lot one at a time, at random, and with equal probability.
Heterogeneity	Nonuniformity in one or more chemicals or physical aspects of interest.
h_i	Heterogeneity carried by particle i.
Homogeneity	Uniformity in all the chemical and physical aspects of interest.
Increments	Individual physical portions that are combined to form a sample.
L	Subscript for lot.
Lot (or population)	The entirety of the material of interest that we wish to sample.
M	Mass (or weight).
N	Number of particles.
Nugget effect	The value obtained by extrapolating the variogram to the vertical axis. It is the mathematical limit of the variation of differences between lagged samples taken closer and closer together in time or space.
Outlier	A value that is somewhat lower or higher than the others.
P	The constant probability of selecting a group of particles in the sample ($= M_S/M_L$).
Preparation error (PE)	The error incurred when the integrity of the sample is not preserved, commonly known as sample handling.
Principle of correct sampling	The principle that every part of the lot has an equal chance of being in the sample, and the integrity of the sample is preserved during and after sampling.
Random sample	A sample obtained so that every part of the lot has an equal chance of being in the sample.
Representative sample	A sample that has the same properties of interest as the lot.

Riffler (riffle splitter)	A tool used to split a sample in half by pouring the lot into chutes emptying into two containers.
S	Subscript for sample.
Sample	The physical material in total that is taken for analysis from the lot.
Sampling dimension	The number of dimensions remaining in the lot if increments are taken across the other dimensions.
Sampling error	The total of all the errors generated when taking, handling, and analyzing a sample.
Static mixer	A device that mixes liquids or gases flowing through it but has no moving parts.
Stream	Stationary or moving material that is longer than it is wide. Solid streams can be sampled by "slicing" across and perpendicular to the length.
Thief probe	A tool used for extracting a vertical core of solid material, as when sampling from a barrel or bag.
Time plot	A graph of measurements plotted against time.
Variance of the fundamental error	Variation in the amount of the constituent of interest obtained from different samples from the same lot.
Variogram	A graph of the variation of samples plotted against the time (or space) lag between the samples.

Chapter 1

Some Views of Sampling

1.1 The goal of good sampling

Why do we sample? In some cases, we sample because we do not have the time, personnel, or money to examine an entire "population" or "lot" (all of the material of interest). In other cases, measuring a property of interest may require destroying the unit, such as in testing the life of a lightbulb or in conducting chemical tests. We may need to characterize the spatial distribution of a contaminant in soil, air, or water in an environmental situation, or we may need to characterize industrial process variation over time. We also use samples[1] for such things as process control, environmental monitoring, and product release.

What kind of sample do we want? If we want to know the "true" overall average of the lot for some characteristic, then we need a sample that is *representative*[2] of the lot. This means we want our sample to be a microcosm of the lot, so that whatever properties we are interested in for the lot, our sample has the same properties. For example, if we are looking for percent impurities in a chemical production batch, then we would like our sample to have the same percent impurities as the entire batch. If we are looking for toxic chemical components in soil, then we would like our sample to have the same composition as the contaminated area of interest (which may not be the entire field).

[1]In this primer, we will use the noun "sample" in a very specific way. *A sample* will mean the physical material in total that is taken for analysis from the lot. It may consist of several "increments" (individual physical portions) that may be combined and measured or measured separately and statistically averaged. The *measurement from the sample* will mean either the chemical or physical measurement of interest on the sample itself (the total material composited from different physical increments) or else the statistical average of the separate chemical or physical measurements on the different physical increments. In other words, the measurement from the sample will mean the value obtained from the sampling procedure or protocol and subsequent chemical and/or physical analysis that is used to estimate the true lot value.

[2]For spatial or temporal characterization, the sample (which may consist of several increments) is expected to be representative only of the immediately neighboring area (in space or time) from which it is taken.

How do we get a representative sample? It's not easy, and we can never guarantee that we have one. But what we will see in the following chapters is that if we follow certain *sampling principles*, then our sample will generally be more representative of the lot than if we do not follow these principles. Therefore, if we want to characterize a lot, however we define it in time or space, then *the goal of good sampling is to follow a sampling protocol that produces a sample whose chemical or physical measurements of interest are*

1. *representative of the entire lot and*
2. *as consistent as possible, theoretically, with other samples that would be obtained if the entire sampling protocol could be repeated.*

Getting representative samples requires using physical sampling techniques that have as little bias as possible in obtaining the total amount of material that makes up the sample. In other words, getting representative samples means *reducing the bias* as much as possible. Getting consistent samples means *reducing the theoretical sampling-to-sampling variation.* That is, if we could go back in time and repeat the entire sampling protocol, and repeat it again, etc., then the variation in the chemical or physical measurement of interest would be "small" between samples from the repeated protocol. Generally, a standard of accuracy and a standard of precision (reproducibility) must be agreed upon by the customer and the supplier.

1.2 Sampling differences

Sampling errors can be costly because they can lead to unnecessary process changes, the analysis of additional samples, or the release to customers of off-spec material (material outside the specifications agreed to with the customer). Further, different methods of sampling can produce different results. For example, Figure 1.1 shows caloric heat energy (BTU) measurements of natural gas over a 12-month period (Welker, 1989).

In one case, a spot sample was taken once during each month. In the other two cases, a continuous sample was composited using a standard 300 cm^3 cylinder and a constant-pressure cylinder. We should not be surprised to see that the BTU values vary each month, regardless of which sampling method is used. It is also apparent that the three sampling methods give quite different values for the same month. For this series of samples, the spot samples are higher in most cases, thus implying that the buyer's energy costs would be greater if that particular sampling method were used. Of course, we don't know what the correct BTU value is, only that the sampling technique is crucial to the result.

1.3 Sampling as scapegoat

When our results differ from what we expect, the discrepancy is often attributed to "sampling error." We recognize the importance of process and laboratory variation (or testing or measurement variation) and what contributes to it, but

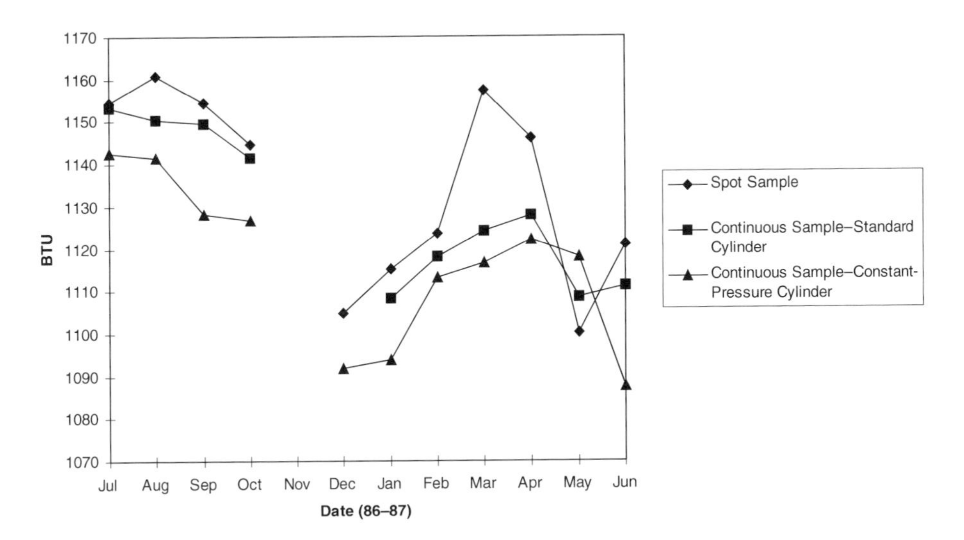

Figure 1.1: Monthly BTU results from three sampling methods.

we tend to lump everything else into one single source: sampling. Moreover, we often use it as the scapegoat for many of the things we can't explain. If we are not aware of sampling problems or if we ignore them, then we cannot separate sampling variation, which may be substantial, from process and laboratory variation. Figure 1.2 shows the proper separation. Frequently, the process or lab variation or both are inflated. In fact, the process personnel may attribute the extra variation that they cannot explain to the laboratory (Figure 1.3), and the laboratory may place it with the process (Figure 1.4). Failure to identify sampling variation as significant can be a source of miscommunication and perhaps even finger-pointing! *Even if we use careful statistical experimentation to separate and measure the contribution of the sampling variation, we rarely take steps to break down further, measure, and reduce the various contributions (components) of this sampling variation.*

1.4 A big problem

Much of what we know about sampling is tied to our experience of specific situations, what has worked in the past, and what has caused problems. Here are some examples of good sampling practices that are probably already familiar.

- If possible, mix the material before sampling.
- Take several increments and composite them to form the sample.
- Collect the sample in a container made of material that will not chemically react with the sample.
- Sample frequently enough to allow for the identification of process cycles.

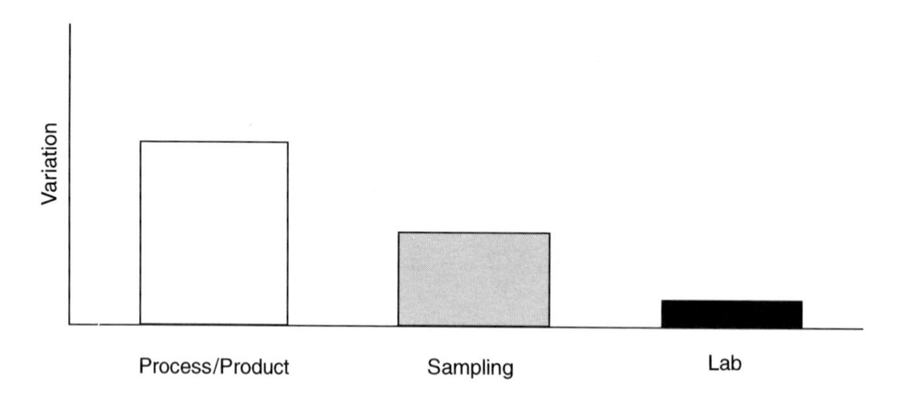

Figure 1.2: Process, sampling, and lab variation separated.

Figure 1.3: Sampling and lab variation not separated.

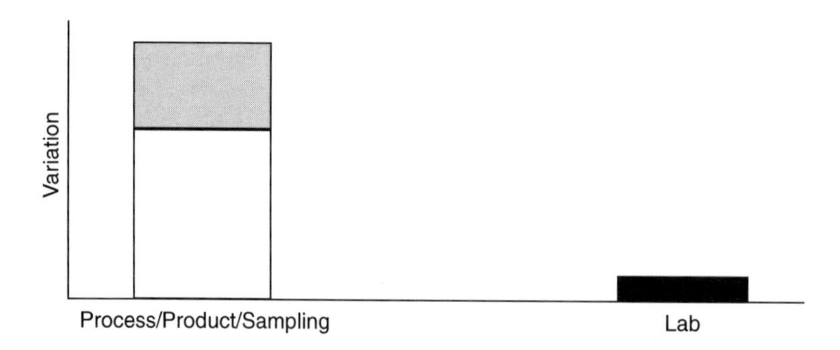

Figure 1.4: Sampling and process variation not separated.

The problem is that there are many more rules than these. How can we possibly list them all for each circumstance we might encounter? How can we remember them? What do we do when new situations arise?

Duncan (1962) touches on many of these problems from a statistical perspective. He examines components of sampling variation in several specific cases and the sampling practices that can affect them, such as compositing, mixing, and determining the size of the sample. He also looks at modeling process samples by accounting for trends and what he calls "short-period" variation. Accuracy and bias are addressed to a limited extent by discussing how to take a sample physically, and he recommends research on the biases of sampling devices. Gy (1992) has done much of this research and answers many of the questions and issues raised by Duncan. Gy addresses all of them by using a structured approach that allows an organized study of the subject.

1.5 A structured approach

Big and complicated problems are most often solved by first breaking them down into component parts and then solving several smaller problems. In the same way, sampling variation can be broken down into component parts so that the different pieces may be examined and the effects of the major contributors reduced. This is the basis of the theory developed by Gy. His theory is not the only one in the literature, but it is structured and addresses all aspects of bulk sampling. He examines sampling variation and bias, provides a method for determining sample size (weight), analyzes process variation and its relation to sampling, examines how to take a sample physically, and introduces the fundamental *principle of correct sampling*. Gy focuses on solids but also examines liquids and gases. The principles he develops apply equally well to all three. He organizes these concepts into seven parts that he calls the seven sampling errors. His theory provides a structure for thinking about sampling and for attacking sampling problems.

Gy's seven sampling errors are sources of sampling variation. Appendix A gives a summary of each of these errors using Gy's terminology. To simplify the presentation for this primer, we group them into three broad categories:

1. material variation;

2. tools and techniques, including sample handling; and

3. process variation.

By learning the basic principles as embodied in Gy's seven sampling errors, we will see that "new" sampling situations are not unique at all. They fall into one or more of these three general categories and thus can be examined using a structured approach.

Chapter 2

The Material: Sampling and Material Variation

2.1 The nature of heterogeneity

Heterogeneous is an adjective defined in *Merriam–Webster's Collegiate Dictionary, Tenth Edition* as "... consisting of dissimilar ingredients or constituents."[†] In essence, heterogeneous means "not all the same" or "not uniform throughout" or "different." Without a doubt, all the materials we sample are heterogeneous, whether they are solids, liquids, gases, or a combination. In other words, there is no such thing as a pure material, either natural or artificial. The presence of *heterogeneity*, the noun corresponding to the adjective heterogeneous, is why physical samples differ and why they generate variation. There are two types of material heterogeneity, and each gives rise to a sampling error: *constitution heterogeneity*[3] (*CH*) and *distribution heterogeneity* (*DH*). From Gy's perspective, "Heterogeneity is seen as the sole source of all sampling errors" (Gy, 1998, p. 24).

Admittedly, these words are a mouthful. But fortunately, they are very intuitive and descriptive. *CH* refers to the differences in the *constitution*, or makeup, of the material: how alike or different the individual particles or molecules are. *DH* refers to differences in how the material is *distributed*: how well mixed or segregated the material is due to density, particle size, or other factors. Each of these two types of heterogeneity gives rise to a sampling error. Together they determine how variable our samples can be and how easy or hard it is to get *consistently representative* samples. Because an understanding and assessment of these two types of heterogeneity are important, we need to examine them in detail.

[3]The term *composition* heterogeneity is more intuitive and could be used instead.

[†]By permission. From *Merriam–Webster's Collegiate ® Dictionary, Tenth Edition* © 2000 by Merriam–Webster, Incorporated.

2.2 Constitution heterogeneity

Constitution heterogeneity (CH) is the variation between *individual* fragments or particles (for solids) or between *individual* molecules (for liquids and gases). "The constitution of a batch of matter is said to be homogeneous when all the constitutive elements making up the batch are strictly identical with one another" with respect to the chemical or physical characteristic of interest (Gy, 1992, p. 49). The batch has a "heterogeneous constitution when its units are not strictly identical" (Pitard, 1993, p. 58).

Gy's *constitutive elements* are the *individual units* that remain *indivisible and unalterable* in a physical, chemical, and time-stable or space-stable context. In particular, solid particles remain in their original state, and ions and molecules in liquids do not react with one another (Gy, 1992, p. 49). In general, *particles* are an agglomeration of more fundamental material. However, in this discussion, we consider a whole "chunk" of material to be a single particle even though it may be broken down into smaller fragments, so long as it meets Gy's definition of a constitutive element during a given sampling stage. Appendix B has Gy's mathematical definition of the CH. The intuitive idea behind the CH is that individual particles or molecules of material vary in the chemical or physical property of interest. For example, soil particles vary in their size, shape, and moisture content. Sheets of writing paper vary, however little, in their dimensions. Gallons of gasoline vary in octane number. An illustration of particle size and shape differences is given in Figure 2.1.

Whether the material is natural or artificial, one sample will vary from the next just because the material is not completely uniform. For naturally occurring materials, the CH is something we are stuck with. For industrial or

Figure 2.1: Constitution heterogeneity. The particles are not uniform.

artificial materials, improving the process might possibly reduce the CH and make a product that is purer or more consistent. The CH gives rise to Gy's *fundamental error* (FE): the difference between what we measure from our sample for the constituent of interest and what the true value of the lot is. One of the challenges we face is getting representative and consistent samples in spite of a large CH. We discuss some ways to do this later in this chapter and also in the discussion of tools and techniques in Chapter 3.

2.3 Distribution heterogeneity

Distribution heterogeneity (DH) is the variation between *groups* of units (fragments, particles, or molecules). A "lot of material has a homogeneous distribution when all the groups of a given size that can be selected within the domain of the lot have a strictly identical average composition; however, the lot has a heterogeneous distribution when these groups are not identical" with respect to the chemical or physical characteristics of interest (Pitard, 1993, p. 58). If the material in the lot is well mixed, then one group of units will have much the same properties as another group. If the material is segregated, such as by particle size, shape, or density, then a group of units in one part of the lot will have different properties than a group in another part of the lot.

The DH is a concern because when we sample bulk solids, liquids, or gases, we do not sample individual units. We sample groups of units, that is, several particles or molecules together. Imagine sampling a gram of powder particle by particle! Figures 2.2 and 2.3 illustrate a large DH for solids and liquids, respectively.

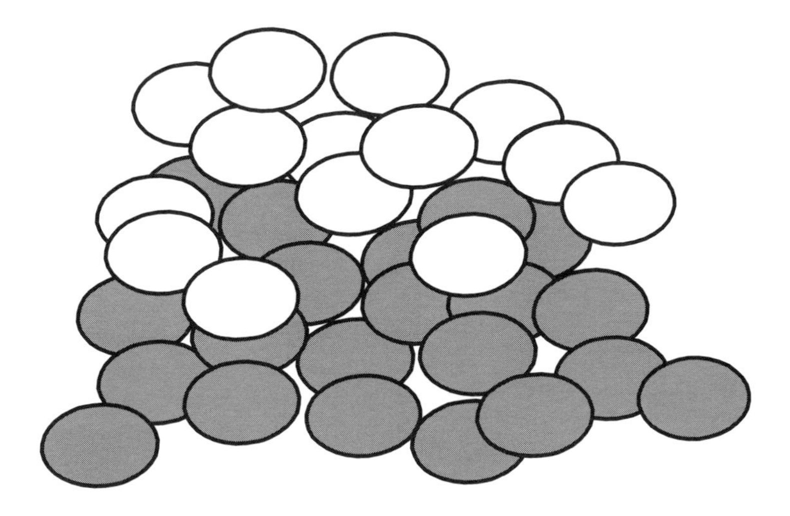

Figure 2.2: Distribution heterogeneity for solids. The particles are not distributed uniformly.

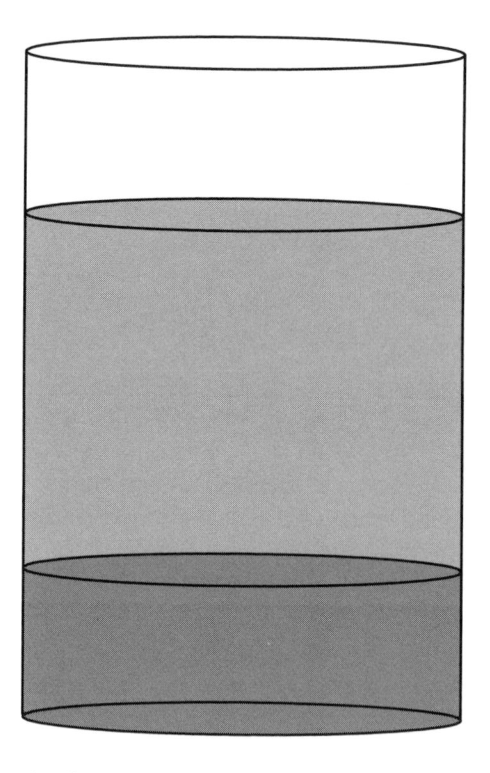

Figure 2.3: Distribution heterogeneity for liquids. The denser, immiscible liquid is at the bottom.

If we could choose the units in our lot one at a time, at random, and with equal probability, then it would not make any difference how they were distributed in the lot. This is the classical "statistical" sampling situation, and the resulting sample is called a *simple random sample.*[4] To illustrate it in a familiar context, suppose we want to select a few people at random from a larger group gathered in a room. We could assign each person a different number. Then we could write each number on a piece of paper, put them in a hat, mix them up, and draw several pieces. (For a large group, random numbers generated from a computer would be easier, of course!) It would make no difference how the people were grouped (distributed) in the room, whether by age, height, income, hair color, political party, or whatever. This method of sampling is the best we can do. That is, this method gives the "best" (unbiased and minimum variance for unbiased estimators) statistical estimate of the true lot content (measurement of interest). In bulk sampling, also, this method is the ideal or theoretical best (Elder et al., 1980), even though it is impossible to sample this way in practice.

[4]Gy (1992, p. 374) calls this the "equiprobabilistic model," where "each of the ... particles composing the lot ... is submitted to the selecting process *independently with a uniform probability* ... of being selected."

In bulk sampling, we do not select the units in our sample one by one but rather in groups. Thus, how the units are distributed makes a difference in the particular sample we get and consequently in the measured amount of the chemical or physical property of interest. On a small scale, the DH gives rise to Gy's *grouping and segregation error* (GSE): the error that arises when we do not sample units one at a time, at random, and with equal probability. If we can reduce the DH, however, we can reduce the sampling-to-sampling variation. That is, we could reduce the theoretical variation we would see if we were able to go back in time, repeat the sampling protocol, and obtain a measurement on each protocol's sample.

2.4 Quantifying heterogeneity

Gy has developed a way to quantify the CH, the DH, and the variance of the FE. By estimating these quantities, we can assess the impact they have on the total sampling variation. This is important if we want to see how best to allocate our resources to improve our sampling results.

Recall that the best sampling we can do is to select units one at a time and at random with equal probability. This results in the minimum sampling error we can hope to attain, since it is the ideal case. We use this ideal case to define the FE as

$$\mathrm{FE} = (c_S - c_L)/c_L, \tag{2.1}$$

where c_S is the measured content (property of interest) in a sample and c_L is the true (unknown) content of the lot. Here, c_S is determined by the ratio (mass of characteristic of interest)/(mass of sample). The FE is the difference between the sample amount and the lot content, measured relative to the lot amount. If the sample is fairly representative, then this difference will be small. Thus, the FE will be small.

Of course, even with probabilistic sampling, the sample almost never has the exact content amount of the property of interest as the lot. Sometimes we overestimate the true lot amount from the sampling and from the chemical or physical analysis, and other times we underestimate it. Fortunately, *on average*, sample results will be *accurate*. This means that if we could take all possible physical samples having the same physical mass by repeating the sampling protocol, where each sample is obtained by selecting the units one at a time and at random with equal probability, the average FE would be zero. Unfortunately, we end up with only one sample (which may be several individual increments or a composite). So this is like the person who is too cold from the waist up and too hot from the waist down. On average, the temperature is right, but the situation is unacceptable.

What we need is some way to get more *representative* samples and get them *consistently*. We need to ensure that the FE is small all the time, or at least as much of the time as possible. In other words, we want the *variance of the FE* to be small. This means that the sample amount c_S never is "too far away" from the lot amount c_L.

For solids, we can use a formula for the variance of the FE to design a sampling protocol to do this. Using experimental results and mathematical approximations, Gy (1992, p. 362) has derived a formula that estimates this variance, Var(FE). It is a function of the mass M_L of the lot, the mass M_S of the sample, and the size, density, and content characteristics of the material:

$$\text{Var(FE)} \cong (1/M_S - 1/M_L) \text{ (size and density factors).} \tag{2.2}$$

This equation can be used to relate sample mass, particle size, particle density, and sampling variation to assess current sampling protocols or to develop new ones. Some details on the derivation and application of this formula are given in Appendix B.

2.5 Reducing the variance of the fundamental error (FE)

We can reduce Var(FE) in several ways to get more consistently representative samples. From (2.2) we see that the sample mass M_S is *inversely* proportional to Var(FE). Thus, assuming the sampling is random, *increasing the quantity M_S of material in the physical sample will reduce Var(FE).* Also, particle size is *directly* proportional to Var(FE). Consequently, *reducing the particle size of the material in the lot* (by grinding, for example) *will reduce Var(FE).*[5] We must take care, however, to preserve the integrity of the sample so that the component of interest is not left in the grinding wheels or smeared on the side of the mortar.

2.6 Relationship between chemical sample size and statistical sample size

Equation (2.2) is similar to a statistical formula relating sampling variation to *statistical sample size.* When statisticians refer to sample size, they mean the *number of units in the physical sample.* When chemists refer to sample size, they mean *the mass, weight, or volume of the physical sample.* This is a basic difference in the use of the terminology as well as in what is being sampled. We discuss here the relationship between the two and how they affect the variation of our estimates.

In classical statistical sampling theory, the sampling units are well defined (Cochran, 1977). Specific units are targeted for sampling. Examples include

[5] It is true that particle size reduction will increase the CH of a lot. However, the number of particles will also increase. In most practical circumstances, this *will decrease the variation* of the measured chemical or physical characteristic of interest if we compare results from one sampling of the lot to another sampling of the same lot. See Appendix B.

people (as in polls), animals (as in wildlife or environmental studies), and parts (as in manufacturing). Equation (2.3) gives an algebraic relationship between the standard deviation (SD) of the population from which an observation comes and the SD of the average of n observations:

(2.3) SD of average of n observations = (SD of population)/SQRT(n),

where SQRT means square root. This relationship holds for almost all common random distributions.

It is easy to see from (2.3) that increasing the statistical sample size n will reduce the SD of the average. Thus, *increasing the number n of units in the total physical sample taken from the lot will reduce the variation of the statistical average of the characteristic of interest.* In other words, sample averages obtained from theoretically repeating the sampling protocol will have the very desirable property of being more *consistent* in the characteristic of interest. *Note that this does not affect any bias that may be present.*

In contrast, when sampling bulk material, the material cannot generally be viewed as a set of distinct units. For example, we sample liquids from tanks, drums, and pipelines, and particulate solids such as ore, powders, and soil. Individual units cannot be identified for sampling. Rather, we must decide on a sample mass M_s or volume: *the chemical sample size.* Further, we must be concerned about whether to composite[6] samples, and, if so, how much to include in each increment of the composite. An additional complication is the restriction on the sample mass that must be used in a chemical analysis due to the method or instrumentation. In fact, a subsample is usually taken in the lab.

Because the sampling unit is no longer easily identified and is sometimes identified only as molecules, we don't have a sample number n of individual units. Rather, we have a sample mass M_s. So we cannot apply (2.3). However, we might expect a similar relationship with sample mass in place of the number n of samples. This is in fact the case, as we illustrate in the following example.

Let us suppose we are sampling solid material and that the sampling units are well defined. We sample n of them. For instance, we might be sampling pellets. Let us also suppose for simplicity that each sampling unit has the same mass M_U. Then the mass M_S of the sample is $n * M_U$. Consequently, we have

$$n = M_S/M_U. \tag{2.4}$$

This relationship is very important for two reasons. First, the statistical sample size n (number of units) and the chemical sample size M_S (mass) are directly proportional. They increase or decrease together when the mass M_U of each unit is fixed. If we increase the mass M_S of the sample, we are also increasing n. That is, we are sampling more units. We saw from (2.3) that *increasing the statistical sample size n* (the number of units in the sample)

[6]Pitard (1993, p. 10) defines a *composite sample* as "A sample made by the reunion of several distinct subsamples. Composite samples are often prepared when it is not economically feasible to analyze a large quantity of individual samples."

reduces the theoretical variation (of the chemical or physical property of interest) from different samples resulting from repeating the sampling protocol. Since n and M_S are proportional, *increasing the chemical sample size* M_S *by increasing the mass of the physical sample will also reduce the theoretical variation between the measurement of samples obtained from repeating the sampling protocol.* In either case, the variance of the average of n units of total weight M_S is the same as the variance of one physical (or composited) sample of weight M_S.

The second reason (2.4) is important is the inverse relationship between n and the denominator M_U, the mass of a single sampling unit. If we grind or crush the particles, we are reducing the particle size M_U and increasing the number n of particles in the sample. Again, from (2.3), increasing n reduces the variation of the sample average. Since n and M_U are inversely proportional, decreasing the individual particle mass M_U by *reducing the particle size will also reduce the sampling variation, even if the sample mass* M_S *is the same.* Appendix B has details.

We can summarize these results as follows.

- For a fixed particle size, increasing the number of individual units in a statistical sample is comparable to increasing the sample weight of a chemical sample.
- For a fixed sample weight, decreasing the individual particle size of material in the lot before sampling has the effect of increasing the number of sampling units.

In both cases, the variation of the sample estimate of the property of interest is reduced. In the latter case, reducing the particle size may not be practical. For example, reducing the particle size of the material in an entire landfill before sampling is impossible. Sometimes, surface soils can be tilled to a certain depth, and this may incidentally reduce the particle size of the top layer. On the other hand, in mining operations, various stages of grinding are commonplace, and sampling can be conducted after particle size reduction.

2.7 Reducing the grouping and segregation error (GSE)

To reduce the effect of segregation on a large scale, we should *mix the entire lot* if possible. For large, immobile lots, such as waste piles, ship cargo, and rail cars, we have to look at alternatives. This is addressed in the discussion of tools and techniques in Chapter 3.

In many cases, shaking will provide good mixing, but there are notable exceptions. Just because material goes through a mixing process does not guarantee that it becomes and remains physically mixed. For example, a container more than about 2/3 full will not provide adequate mixing. Solid particles that differ in size, density, and shape are especially susceptible to poor mixing, and shaking or rotation may actually increase segregation (Leutwyler, 1993, and Studt,

1995). Even after mixing, solid granules may resegregate during handling and storage (Johanson, 1978). Immiscible liquids of different densities do not mix well and may separate rapidly. Solids that do not go into solution may settle quickly or float, depending on whether they are heavier or lighter than the liquid. In pipes, the degree of mixing of certain liquids can be affected by the flow rate (ASTM, 1982). Thus, mixing and its opposite, segregation, are transient phenomena. As such, any variation in the distribution of the material imposes a bias that varies in space and time.

If possible, the degree of mixing should be ascertained. If solid particles are easily differentiated by size, shape, or color, for example, or different liquids by color or density, it is easy to see the effects of mixing. In addition, an experiment can be performed to measure the effectiveness of mixing in the following way. Conduct the usual amount of mixing. Then separately analyze portions of material from different parts of the lot for the component of interest and compare the results. If the variation in results from the different portions is acceptable, then the mixing is adequate.

To reduce the effect of sampling groups of units rather than individual units one at a time, we should take several small increments and combine (composite) them to get our sample. The smaller the groups we take to put into our sample, the closer we are to the theoretical best: random sampling unit by unit. The worst thing we can do is to take one sample in one big group to estimate the critical content of the lot, even if the lot is a very small area of soil or a very small amount of material, such as in the lab. This is easy to see in an extreme case, where the lot is completely segregated by some property of interest. A sample consisting of a single group would be restricted to one part of the lot and the sample would not be representative of the lot. Taking several increments (or portions) randomly allows sampling from different parts of the lot and therefore produces a more representative, combined sample.

2.8 Summary

There are several ways to reduce the influence of the constitution and distribution heterogeneity (CH and DH) on the sampling variation. These are things we can do to increase our chances of getting representative samples and getting them consistently.

1. *Increase the mass of the total physical sample.* This reduces the theoretical sampling variation that results from the inherent heterogeneity of the material. From an intuitive perspective, the more units, particles, or molecules in a lot that we select (randomly) to be part of the sample, the better idea we have about the true lot properties.

2. *Collect several (random) increments from the lot and combine them to form the sample.* This protects us against the possibility that the lot is not well mixed.

3. *For solids sampling, grind the particles in the lot before sampling.* Particle size is directly proportional to the sampling variation. Thus, reducing the particle size will reduce the theoretical sampling variation, even when the sample mass is not increased.

4. *Mix the material.* This reduces the lot segregation and consequently the inflationary effect on sampling variation resulting from selecting groups rather than individual units.

Chapter 3

The Tools and Techniques: Sampling, Sample Collection, and Sample Handling

3.1 Representative sampling

As discussed in Chapter 1, we would like to obtain samples that are *representative* of the entire lot. That way we will know exactly (or at least within the analytical measurement) the values of the properties of interest. From our experience, however, we know that these values vary from sample to sample. Thus, the samples cannot all be representative. How can we get representative samples, or at least approximately representative samples?

The key is to apply the statistical principle of random sampling. We saw in Chapter 2 that the sampling of individual units at random minimized the sample-to-sample variation. Random samples may seem to be taken in an arbitrary and unorganized fashion. But, in fact, they are more representative, and consistently so, in repeated sampling situations (day in and day out) than samples obtained any other way. Actually, we all practice random sampling in simple situations, as we will illustrate below. The real trick is applying this principle to solids, liquids, and gases, where individual units are rarely available for selection.

3.2 Random sampling

When we have to select a subset from a larger lot, we are always told to take a *random sample*. We have all seen this, for example, when we draw names out of a hat to determine teams for parlor games, flip a coin to make either/or choices, or draw straws to select one person out of a group.

Why do we do this? Does it make sense? In the cases above with which we are familiar, we have a sense of *fairness*, of unbiasedness. And *unbiased* is the word used by statisticians to describe the fact that the average of estimates obtained from many random samples will equal the value of the entire lot. When we sample randomly, the laws of probability apply, meaning that the odds are in our favor of getting a representative sample, though there is no guarantee. This also means that when we get an estimate of a value of the entire lot based on examining a random sample, we can calculate an estimate of the *statistical sampling error*. Thus, by taking a random sample, we not only have a statistically unbiased estimate, but we also have an idea of how good or bad that estimate is. For example, in polls taken to determine voter preferences of political candidates and issues, the results are stated as a percent with typically a 2% or 3% error. Clearly, the smaller the estimation error, the more faith we have in the results obtained. Thus, knowing the error of our estimate is valuable information.

Taking a random sample therefore gives us

1. *a statistically unbiased result and*

2. *a statistical estimate of the precision of the result.*

Random sampling is used to ensure as much as possible that every unit in the lot has an equal chance of being in the sample. Seen in this way, it is a very basic idea that is easy to justify intuitively, understand theoretically, and accept in practice. The problem is implementing it. In the simple cases mentioned above, we must make sure that the names in the hat are well mixed, that the coin is fair, and that all the straws look the same to those selecting them.

The statistical theory of random sampling is based on lots (or populations) with well-defined units, such as a population of people or manufactured parts (Cochran, 1977). How do we apply the idea of random sampling to amorphous material? As we saw in Chapter 2, in the case of liquid and gas sampling, a unit is not even defined, except perhaps as an individual molecule. In the case of solids, the units may be so numerous as to make numbering impossible. Thus, the technique of assigning each unit in the lot a different number and randomly selecting a subset will not work. In addition, some parts of the lot may be inaccessible, such as when sampling from a pile or a rail car. If we are trying to sample a powder, how do we make sure that every fine particle has an equal chance of being in the sample? We can't number each fine particle, so what does random sampling mean in this case?

3.3 The principle of correct sampling

Duncan (1962, p. 330) has stated that "Randomness in the selection of original increments is indeed the number one problem in the sampling of bulk material."

Fortunately, Gy (1998, p. 28) has developed a principle for randomness to apply to bulk sampling and has found ways to implement it.

The principle of correct sampling for bulk solids, liquids, and gases

- *Every part of the lot has an equal chance of being in the sample and*
- *the integrity of the sample is preserved during and after sampling.*

This principle is the guide for the random sampling of amorphous material. In the first part, we have to make sure that we

1. *theoretically define the sample* we will take so that it follows this principle and
2. *physically obtain the sample* we have correctly defined.[7]

Failure to carry out these two steps properly contributes most often to bias in sampling. If part of the lot is inaccessible or the sampling tool cannot take the sample we have correctly defined, then getting a random sample is not possible. In the second part of the principle, we must ensure that we

3. *preserve the integrity of the sample* while it is being taken and also between the time it is taken and the time it is analyzed.[8]

Oxidation, abrasion, and evaporation are examples of improper sample handling because the sample integrity is not preserved. Getting a representative sample is useless if the value of the property of interest changes during transport. Then it is no longer representative.

Later in this chapter, we will examine these three steps in detail to understand how the principle of correct sampling can be followed or violated.

3.4 Grab sampling: A common but poor practice

The most common error in *defining the sample* is grab sampling. Gy defines grab sampling as "the art of collecting increments from the most accessible part of the batch." The more heterogeneous the lot in its constitution or distribution (CH or DH), the less representative a grab sample will be. Examples of grab sampling include sampling from the top of a pile, the bottom of a drum using a spigot, the side of a conveyor belt, or the bottom of a pipe. The latter two are illustrated in Figure 3.1.

[7] Gy uses the terms *delimitation* (defining the sample, including its boundaries) and *extraction* (obtaining or recovering the sample).

[8] Gy refers to improper sample handling as *preparation error*.

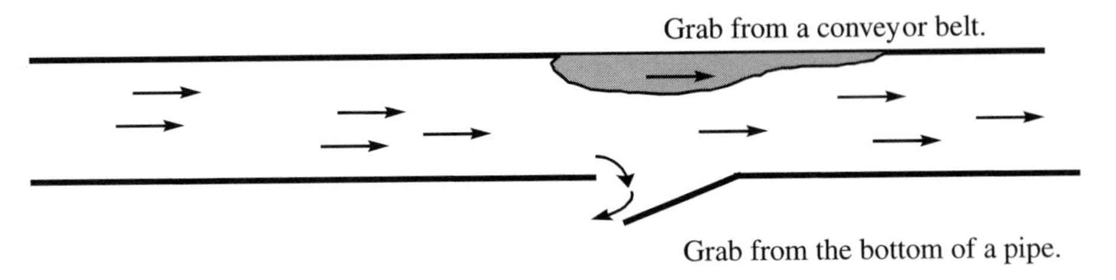

Figure 3.1: Examples of grab sampling from the side of a conveyor belt or from the bottom of a pipe.

Grab sampling does not follow the principle of correct sampling since certain parts of the lot have no chance of being in the sample. Thus, our estimate of the amount of the constituent of interest may be biased, and we cannot calculate a statistical error for it. In other words, not only have we reduced our chances of getting a representative sample, but we also have no idea how bad it is! Autosamplers that penetrate only the edge of a flowing stream, as in Figure 3.1, are commonplace, but they are just glorified grab samplers.

The most common error in *physically obtaining the sample* occurs when the sampling tool cannot take the sample that has been properly defined. For example, an extraction error is produced by a retractable cross-stream sampler that goes only partway across the stream before returning to its idle position.[9] This is in effect a grab sample since only one side of the stream is sampled. Any segregation across the stream will produce a biased sample.

Two reasons why grab sampling is so popular are that it is easy to perform and that its consequences are not realized. For grab sampling to be correct, the lot must be perfectly homogeneous. Since we know this is never the case, then the lot must be and remain well mixed. However, this is not dependable either, so a grab sample will always be biased. Unfortunately, grab sampling is often seen as the only option in cases where the lot is stationary and some of the material is inaccessible, such as with piles, drums, and tanks, and even with material conveyed in closed pipes. However, as will be seen, the lot can be viewed in a different way that perhaps improves the sampling technique. This perspective is based on the idea of sampling dimension.

3.5 A crucial determination: The sampling dimension of the lot

Since we live in a three-dimensional world, we may think of all lots as being three-dimensional, actually four-dimensional if we include time. We will see, however, that by defining increments and samples in a certain way, some of the

[9]By *stream* we mean stationary or moving material (solids, liquids, gases, or discrete units) that is longer than it is wide and can be sampled by "slicing" across and perpendicular to the length.

dimensions may be ignored. The number of dimensions remaining is called the *sampling dimension*. Determining the sampling dimension is crucial in deciding whether correct samples can be taken. Reducing the sampling dimension, either in fact or artificially through sampling techniques, generally helps reduce sampling error. In this section, we examine the different sampling dimensions and how well the principle of correct sampling can be applied in each case.

Zero-dimensional sampling

Consider the cluster of 27 blocks shown in Figure 3.2.

To take a random sample of 2 blocks, we could number the blocks in the cluster from 1 to 27. Then we could generate 2 random numbers from a computer, or, alternatively, number 27 pieces of paper, mix them, and draw 2 at random. We select the 2 blocks with those numbers as our sample. Since all the blocks are easily accessible, this procedure is fairly straightforward, and we have followed the principle of correct sampling.

- Every part (block) of the lot has an equal chance of being in the sample since we are choosing the numbers randomly.
- We have no difficulty physically extracting the two blocks chosen.
- We have no sample-handling issues.

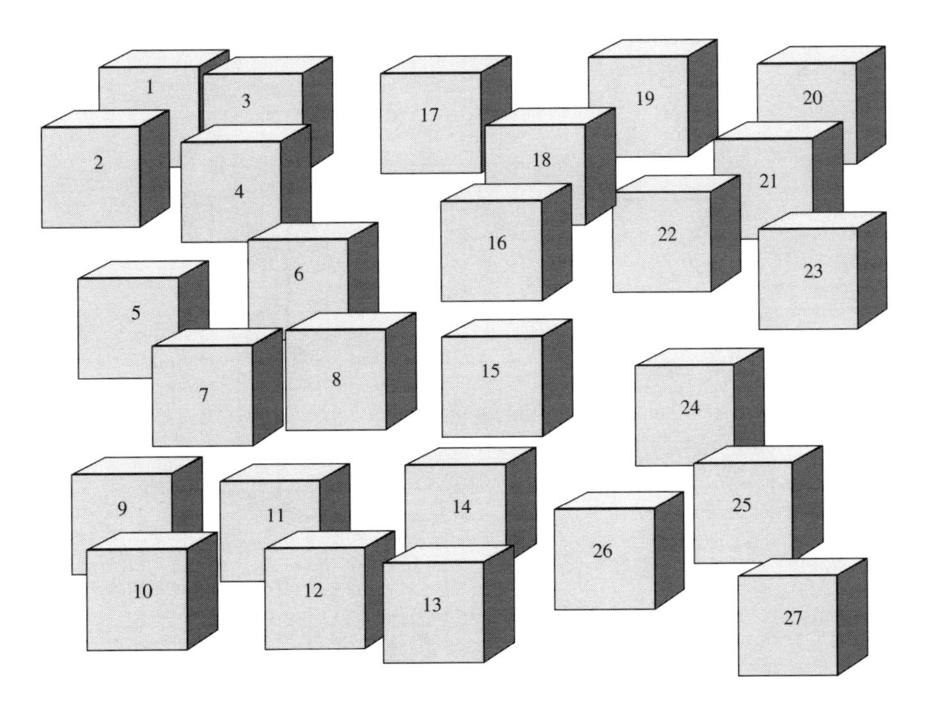

Figure 3.2: 27 blocks numbered individually.

Gy refers to this sampling situation as *zero-dimensional.* As we saw in Chapter 2, he develops a model for the variance of the FE and bases it on statistical sampling theory, where all the sampling units are discrete and well defined. *As long as all units are accessible*, we can easily define and extract the units for the sample. Adhering to the principle of correct sampling is thus straightforward.

In practice, however, there are many situations where actually *obtaining* the units identified for the sample is not logistically feasible. This happens, for example, when a large number of bags or drums is stacked vertically and pushed together in a warehouse for storage. This leads to a common sampling situation, discussed next.

Three-dimensional sampling

Now suppose that the 27 blocks are arranged in a large $3 \times 3 \times 3$ cube as in Figure 3.3.

We could number the 27 blocks and determine the 2 in the sample by random numbers as before. However, actually obtaining the 2 chosen blocks is a little trickier since our sample may consist of blocks in the middle and bottom layers as shown. With only 27 blocks, this is not a big problem, but it illustrates the difficulty of sampling from a three-dimensional lot. If we had a large number of bags of product that were stacked and layered, then pulling out bags selected for the sample would be a logistical nightmare. Thus, even with discrete units, 3-dimensional sampling can be difficult in practice.

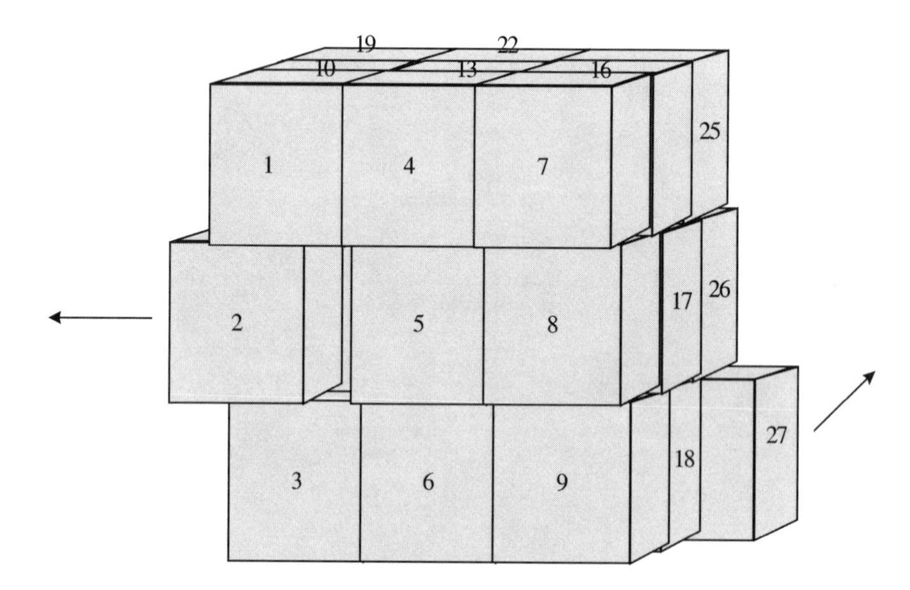

Figure 3.3: 27 numbered blocks arranged in a cube.

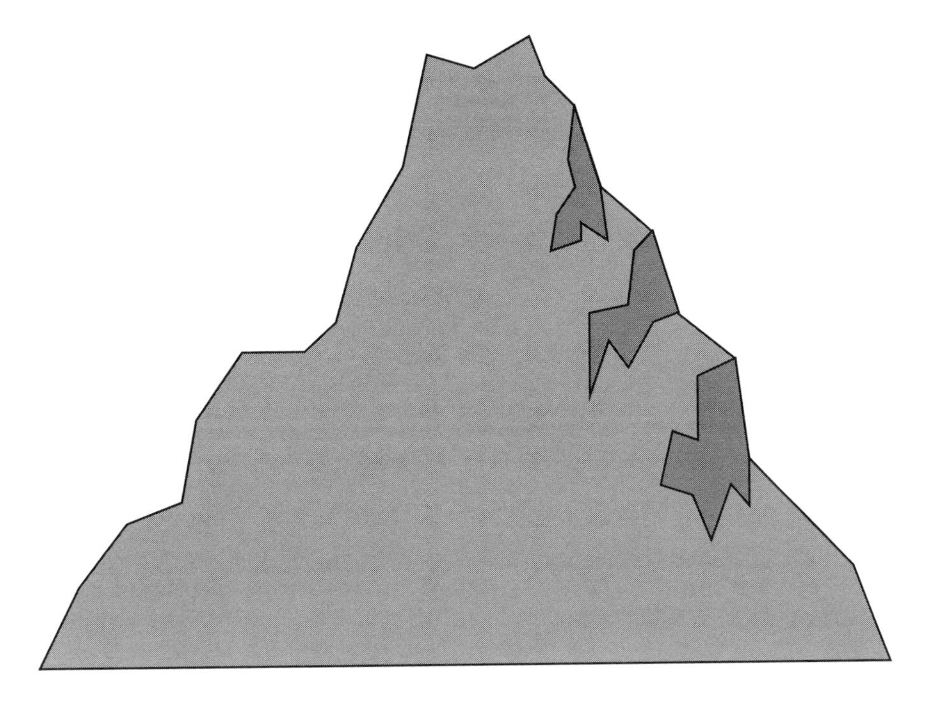

Figure 3.4: A pile of solid particles.

If we generalize this situation to a three-dimensional pile of solid particles, as shown in Figure 3.4, we couldn't number every particle. Even if we could, we could not take it apart as we do the blocks to get a particular particle. We couldn't even find it, because as soon as we started to go into the pile, we would disturb it and thus lose track of which particles are supposed to be in the sample and which are not.

Gy concludes that except in the case of nonviscous liquids, the principle of correct sampling cannot be followed in three-dimensional sampling situations. Thus, three-dimensional sampling situations should be avoided if possible, and we discuss next a way to do this that can be applied in many instances.

Two-dimensional sampling

Suppose we define the sampling unit as a vertical stack of three blocks, and we select two *stacks* as our sample, as shown in Figure 3.5.

We can still take a random sample, but now we have 9 rather than 27 sampling units. We number the sampling units (stacks) 1 to 9 in *2 dimensions* and choose 2 stacks at random. Notice that we ignore the 3rd dimension (depth) in the numbering scheme but sample across this dimension in the definition and extraction steps. Even if a stack selected randomly for the sample were in the middle (#5) as shown, we might be able to pull out the whole stack. This illustration is similar to that of taking a sample of solid particles from a drum

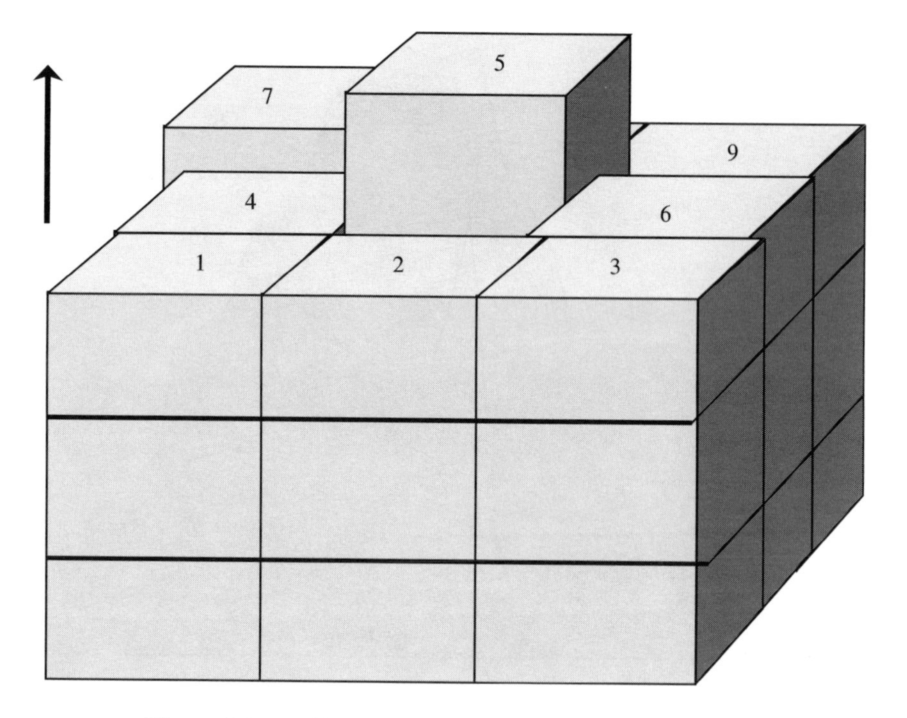

Figure 3.5: 27 blocks arranged in 9 numbered stacks.

or rail car with a thief probe, shown in Figure 3.6, where a vertical core is the sample. Limitations of this tool are discussed later in this chapter in Section 3.6.

By grouping units vertically and defining our sample vertically, we have artificially eliminated one dimension, the vertical dimension, of the lot. *So the sampling dimension is two.* That is, we define our random numbers in a two-dimensional plane and sample across the third dimension. This sampling technique protects against the likely situation that the lot is segregated or layered by depth. Because of possible vertical heterogeneity by depth, the vertical sampling core should be a cylinder, ensuring that the same amount of material is taken at the top, bottom, and in between. *A cylinder thus describes the correct geometry for a sampling dimension of two.*

By reducing the sampling dimension from three to two, we can improve our chances of taking a good sample. However, two-dimensional sampling still presents practical problems in terms of defining a probabilistic sample and then actually extracting the sample defined. We can further improve our chances of implementing the principle of correct sampling by eliminating yet another dimension, if possible.

One-dimensional sampling

To continue with the illustration of the blocks, we now group them in 2 dimensions and sample across the 3rd. We do this by defining the sampling unit

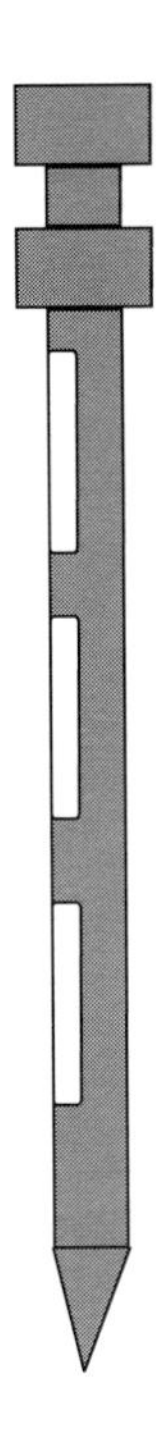

Figure 3.6: Thief probe for solids sampling.

as 9 blocks "in a plane," that is, 3 consecutive stacks of 3. In this case, the sampling unit is 9 blocks rather than 3 blocks, and we choose 1 of the 3 9-block groups at random for our sample. Sample extraction is now fairly straightforward. As shown in Figure 3.7, we simply push out the 9-block slice that makes up the sample, based on which random number is generated. By grouping the units in 2 dimensions and numbering them in the other dimension, we have *one-dimensional sampling.*

This idea generalizes nicely to a pile of solid particles, as long as the pile is fairly flat and as long as we can cut across the entire width of the pile. As shown in Figure 3.8, we can take several slices or cuts across the pile, which would be our increments, and then combine them into a sample.

This technique is applicable in the field on a large scale with a front loader as well as in the lab on a small scale with a scoop. We still have particle disturbance as we cut through the pile, but not as much as if we had tried to take a vertical core as in two-dimensional sampling.

The correct geometry for one-dimensional sampling is a "slice" across the other two dimensions of the material. This applies whether the material is moving or not. A tool that collects a sample by going across a moving stream must maintain a constant speed. If it moves more slowly at first then picks up speed, it will collect more material on the side where it starts. A tool that is placed vertically to get a "cut" across a nonmoving stream must have parallel sides. Otherwise, more material will be collected where the sides are wider.

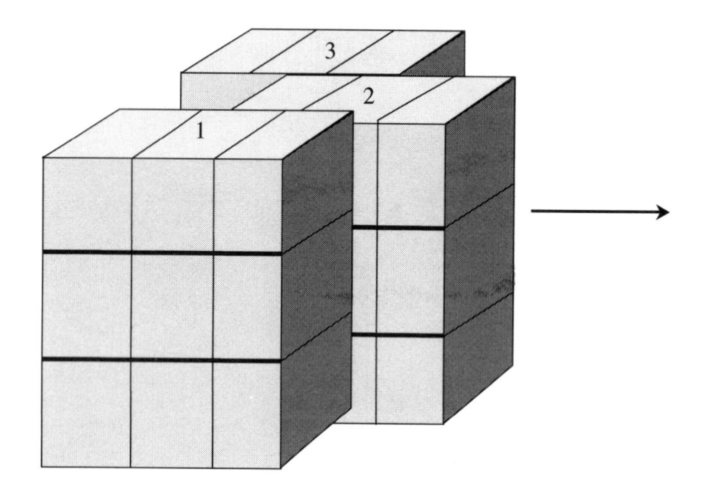

Figure 3.7: 27 blocks arranged and numbered in 3 planes.

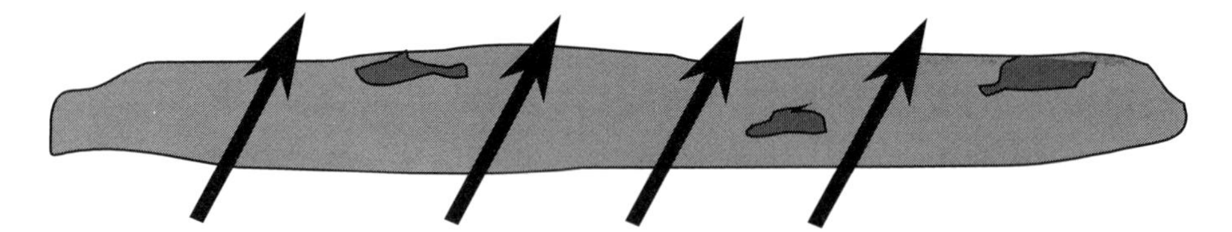

Figure 3.8: Slicing across a pile.

For one-dimensional sampling, conditions must be present to ensure an equal probability of being in the sample for all particles across the whole "width" of the stream. For most solids sampling in commercial processes, a one-dimensional sample is practical. It also provides the best chance of following the principal of correct sampling. For liquids and gases, one-dimensional sampling is more difficult, but in the next section, we discuss ways to minimize the sampling errors that remain.

We are fully aware that one-dimensional cross-stream sampling is not always possible in practice, especially for liquids and gases. However, it is important to look for alternative sampling points in the process that will allow reduction of the sampling dimension. We will then be better able to follow the principle of correct sampling and thus improve our chances of getting good, representative samples.

3.6 Defining and extracting the sample

To get a correct sample, we must define the sample using the correct geometry for the sampling dimension we have produced or been given. We must then use a tool that will extract the sample we have defined. As we have seen, practical situations are not nearly as easy as sampling blocks. Consequently, we may not be able to reduce the sampling dimension. Even if we do, we still may compromise the principle of correct sampling. In this section we examine several sampling situations and the tools that might be used. Both advantages and shortcomings will be discussed.

A common tool used to sample bulk solids is a thief probe, shown earlier in Figure 3.6. When the sampling dimension is two, a thief can be used to sample across the third dimension. It is often employed for sampling a vertical core from large bags or drums. For taking a horizontal cross section from a small bag, the point is used to punch through the outside packaging and push through the material. A knob at one end is used to open and close a set of windows along the length of the cylindrical shaft. The pointed end allows penetration of the material. To use the thief, the windows are closed and the thief is pushed through the material. The windows are opened to allow material to flow in and then closed before pulling out the thief. For loosely packed solids, this tool works fairly well.

The thief probe has some limitations, however, which can compromise the principle of correct sampling. With fine powders, the windows may leak if not constructed properly and fill from the top before the thief gets to the bottom. On the other hand, material may not fall into the windows after they are opened. Also, the thief may be difficult to push all the way through the material. Even if the point gets all the way to the bottom, the bottom will never have any material in the sample. This is not a true vertical core, and vertically segregated material results in a biased sample. Gy (1992, p. 230) gives an example of huge monetary losses by a sugar beet refinery that relied on a type of thief to sample incoming loads of sugar beets. The refiner was unaware of flint pebbles on the truck bottom put there by the producers as waste but paid for as product by the refiner.

For liquids, there are several ways to apply the principle of correct sampling across the vertical dimension. A tool that uses the same idea as the thief is the coliwasa (COmposite LIquid WAste SAmpler), shown in Figure 3.9. They come in various lengths and diameters, depending on the depth and size of the container lot. This hollow column is lowered slowly through the liquid until the desired depth is reached. Then the top and bottom are plugged to retain a core.

ASTM D 4057 (1981) describes a way to get a vertical top-to-bottom sample of petroleum products from a storage tank. A stoppered bottle is dropped vertically all the way to the bottom. Then it is unstoppered and pulled up at such a rate that the bottle is 3/4 full as it emerges from the top. Unfortunately, this is difficult even for a seasoned practitioner. Another way to sample vertically is to take samples from the top, middle, and bottom, also discussed in ASTM D 4057. A stoppered bottle is lowered to the desired depth, the stopper is pulled, the bottle is allowed to fill, and the bottle is raised. These latter samples are easier to obtain because they require little expertise. They also incur less extraction error. They do not give a full vertical cross section but give some representation of the different depths.

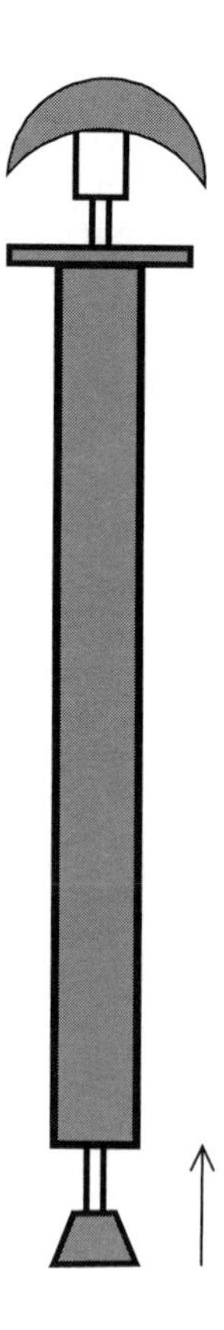

Figure 3.9: Coliwasa takes a vertical core from a liquid.

One-dimensional sampling is preferable if it can be arranged. Still, problems present themselves that are not obvious. Slicing across the pile in Figure 3.8 often results in an extraction error. This error can be minimized, however, if a proper tool is used. For example, the tool must be large enough to hold all the material in an entire slice. Otherwise, material at the end of the slice that should be in the sample will be left out. The sample will thus be biased in favor of material taken early. A sampling tool that is round at the bottom will not take as much material from the bottom and will thus be biased in favor of material on the top. A correct sampling tool in this case has flat sides that are perpendicular to a flat bottom, as illustrated with the hand scoop in Figure 3.10.

Even with proper sample definition and extraction, a sample-handling error can occur. For example, static electricity may be generated when sampling a fine powder. Powder on the *outside* of the scoop does not belong in the sample but may fall into the sample container if the scoop is tapped against the container wall to remove clinging particles. Powder on the *inside* of the scoop belongs in the sample but may cling and remain there when transferring to the sample

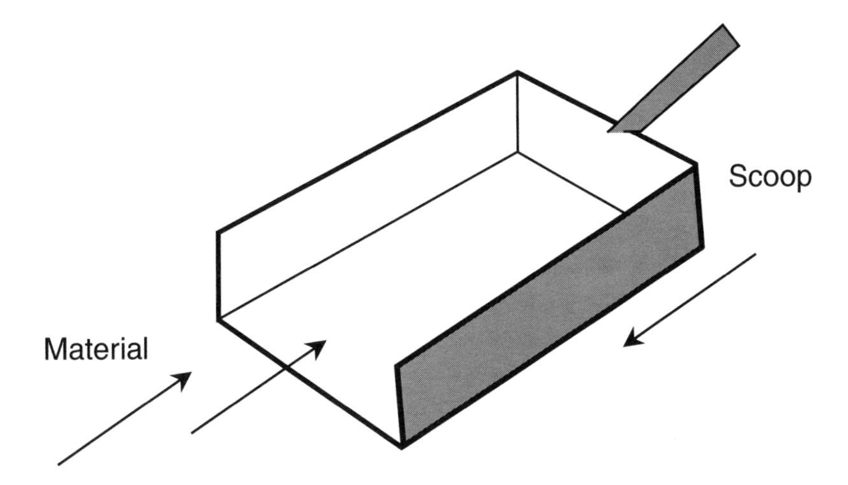

Figure 3.10: Correct sampling tool for sampling across a solid pile.

container. In either case, the sample integrity is compromised. If the scoop is not cleaned in between samples, then particles from one batch contaminate samples from the next. If we are looking at particle size distribution, then the percent of fines will be biased low if some remain on the scoop, and the percent of larger particles will be biased high.

Riffle splitters (rifflers) are used frequently on the plant floor and in labs to mix as well as to subsample solid particles. An example is shown in Figure 3.11.

Let's look at this tool in detail. Material is poured across the riffler in a direction parallel to the channels. Adjacent channels slant in opposite directions. The material flows down the channels into receiving pans on each side. With roughly half the material in each pan, one pan is chosen at random for the sample. The procedure can be repeated several times to get subsamples.

Notice that several increments are used to form the sample, thus reducing the grouping and segregation error (GSE). This same idea was illustrated in Figure 3.8 when slicing across a pile. In that case, the material was stationary and the sampling tool moved across. In this case, the sampling tool (riffler) is stationary and the material is moving. In both cases, several cross-stream increments are combined to form the sample. While using a riffler looks straightforward, Gy points out several ways that it can be easily misused and that will result in biased samples. Consequently, the following rules should be observed.

1. The material must be fed from the feeder pan (not a receiving pan) with material spread evenly across its width.

2. The edge of the feeder pan must be perpendicular to the direction of the chutes.

3. The open end of the feeder pan must cover the set of chutes exactly.

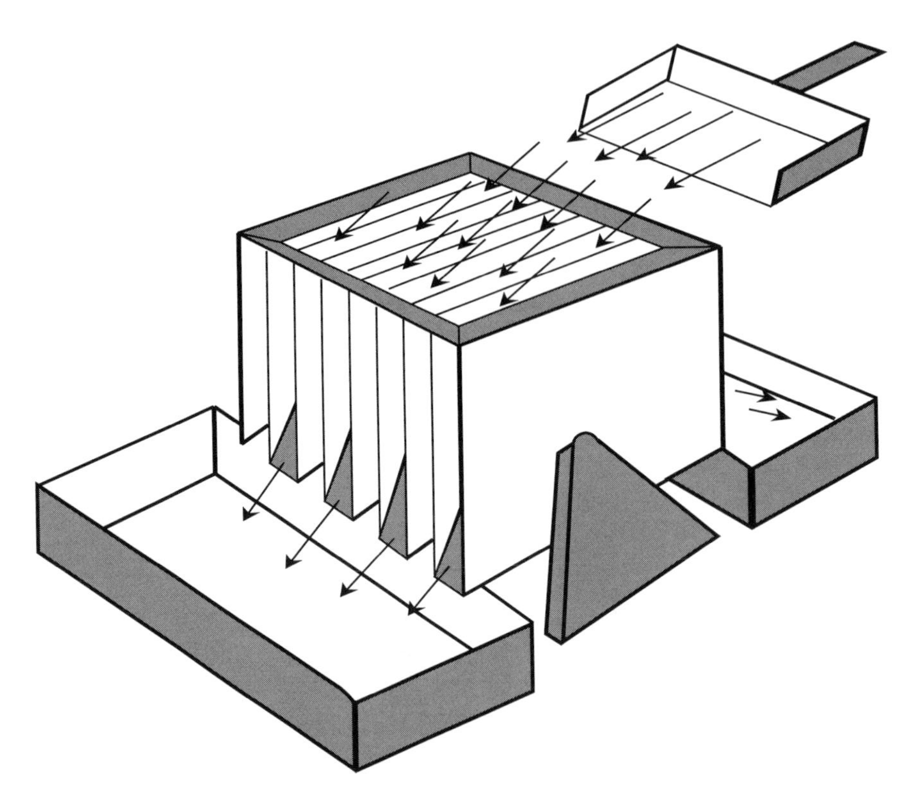

Figure 3.11: Typical riffle splitter used for subsampling solid particles.

4. The material must be gradually fed into the middle of the chutes.

5. The receiving pans must be large enough to hold all the material to be split.

6. The material must not flow out of the channels so fast that it overshoots the receiving pans.

7. The material must not bounce out of the receiving pans.

8. The selected sample must be chosen at random.

Rules 5–7 address sample handling and may be difficult to follow with this particular riffler, especially when sampling pellets or beads. These problems can be avoided, however, by using receiving pans that are high enough on the back and sides to completely cover the channel outlets (Smith and James, 1981, p. 94).

A different kind of riffler that generally produces better samples, that is, samples that are more accurate and precise and thus more representative, is a spinning riffler (Allen and Kahn, 1970). Material is automatically fed by gravity

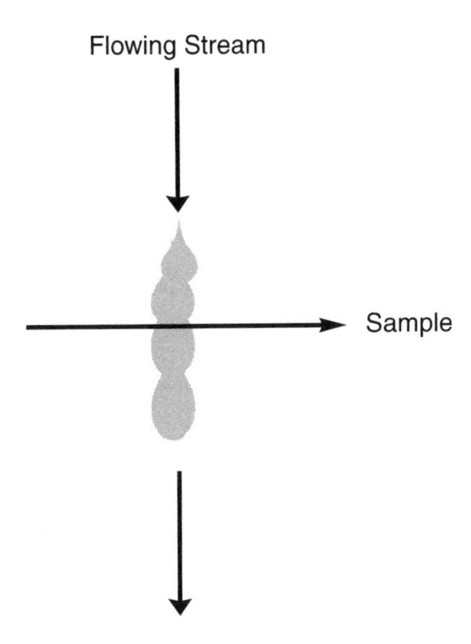

Figure 3.12: Taking a cross-stream sample from a flowing stream.

into separate receiving containers, usually six or more, which rotate in a circle under the feeder. However, just because it is automatic and more complicated does not guarantee that it is unbiased (Wallace and Kratochvil, 1985).

We cannot always take a one-dimensional slice across a three-dimensional lot because the material may be in a container. However, we may be able *to sample the material as a flowing stream before it becomes a stationary lot.* In the example in Figure 3.12, we can sample material as it flows in "one dimension" out of the loading arm, before it becomes three-dimensional in the truck or rail car. This is not practical, however, when the material flows very fast and splashes or is hazardous.

Solids moving along a conveyor belt provide the opportunity for correct, one-dimensional, cross-stream sampling before the material collects as a three-dimensional lot where the principle of correct sampling cannot be followed. In both of these cases, we also have the fourth dimension, time, where sampling frequency and process variation add complicating elements. These will be explored in detail in Chapter 4.

One-dimensional liquid and gas sampling presents special problems. Obtaining a cross-stream sample is difficult if not impossible, and the lot itself is disturbed when trying to take the sample. Liquids coming out of a pipe or hose may flow so fast that moving the sampler across the stream sprays the liquid all around. We must take special safety precautions for liquids and gases that are flammable or whose fumes are toxic. This precludes taking a correct sample. Appendix F discusses safety aspects in various sampling situations. Preserving the sample integrity is particularly difficult in some cases. Examples include evaporation of the component of interest, reaction of the liquid with

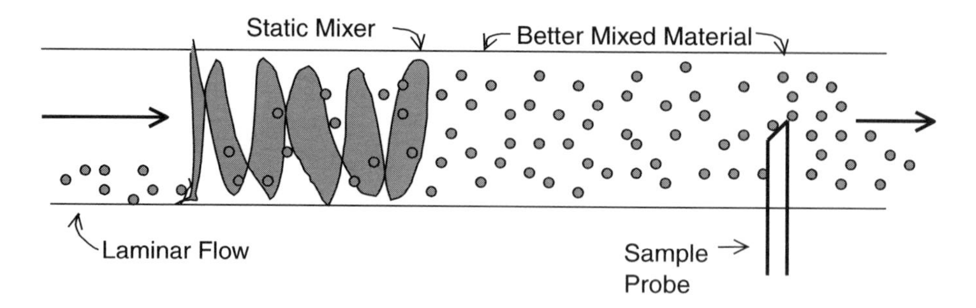

Figure 3.13: A static mixer conditions the stream before sampling (top view with sample probe coming off the side).

the container material, and changes in composition due to exposure to light or air.

No one has yet figured out how to take a correct sample for liquids and gases conveyed in pipes. Because of laminar flow with some liquids, segregation across the stream results in biased samples obtained from on-line analyzers that take a grab sample from the side of the stream. Oil and water mixtures are particularly susceptible to this. If we cannot take a correct cross-stream sample, then we must mix the material well to reduce the bias resulting from the GSE. Unfortunately, some of the techniques thought to be good for mixing, such as those in ASTM D 4177 (1982), do not apply in all cases. Welker (1984) has experimented with liquids moving through plastic and glass pipes. By observing liquid behavior through these transparent pipes, we can confirm the following.

- Increased stream velocity does not necessarily cause mixing.
- A sample loop does not necessarily cause mixing.
- A strainer is not a mixer.

Static mixers[10] like the one shown in Figure 3.13[11] are useful for conditioning a liquid stream prior to sampling, reducing the GSE.

ASTM D 4177 (1982) has detailed recommendations for sampling petroleum and petroleum products from pipes, such as placing probes on a vertical section in the middle of the pipe after mixing. It also includes information on examining the adequacy of the probe location, the distance before or after a bend in the pipe. Welker (1997) warns that the inlet to the sample collection head must be large enough to inhibit the diversion of sample droplets. While neither of these sources considers the principle of correct sampling, both address minimizing the GSE to increase the chance of getting a representative sample.

[10]A *static mixer* is a "mixing device that has no moving parts. The kinetic energy of the moving fluid provides power required for mixing." ASTM D 4177 (1982).

[11]For a horizontal stream, the sample probe should come out of the side for liquids and off the top for gases.

We can apply the idea of cross-stream sampling even if the sampling units are discrete. Consider the situation discussed earlier, where we needed to select randomly several bags of product from a large lot. If the bags are not easily accessible because of the way they are stored, then statistical techniques can be used to obtain a random sample of bags *as they are filled at the end of the process.* The sequence of bags is sampled as a one-dimensional "stream" before they are stacked and stored, that is, before they become a three-dimensional lot. We give details of one such method in Appendix C.

3.7 Sample handling

The second part of the principle of correct sampling requires that the integrity of the sample be preserved. Carelessness and lack of training are the main reasons this fails in practice. As we have illustrated, collecting representative samples consistently requires correct use of the proper sampling equipment. We have also seen how easily these efforts can be neutralized if the property of interest changes somehow as the sample is taken or as nonsampling activities such as transfer, transport, storage, or particle size reduction take place. *The best resource to help with protocols, tools, and containers that will ensure preserving the sample integrity is a chemist.* Many sample-handling errors have associated chemical issues, and chemists are well versed in sample handling from their lab experience. Gy identifies several categories of sample-handling errors, some of which we have addressed in previous examples. In this discussion, we put them into three groups for simplicity:

- contamination,
- loss, and
- unintentional mistakes or intentional tampering.

Contamination

Sample contamination occurs when extraneous material or molecules are added to the sample after the sample is taken but before the chemical or physical analysis. This very common problem occurs with solids, liquids, and gases. One simple source of contamination occurs because the sampling tool or sampling equipment is not cleaned. Material from one batch or even from another product is left in the equipment and becomes mixed with material in the next sample. It is ludicrous for production personnel to sample solids day in and day out with the same hand scoop and never clean it, while at the same time the lab personnel go to great lengths to ensure the integrity of the sample they receive by using, for example, sterile vials, nitrogen purges, and refrigeration.

Contamination occurs when the sample lines are not purged (flushed) at least the length of the line to avoid getting old material, material that was drawn into the line when the last sample was taken. Sampling tools or sampling containers that have trace amounts of the property of interest can contaminate the sample. For example, metal scoops, rifflers, and holding trays should not be used to sample or store material that will be measured for certain metals. Contamination of the sample material will result in measurements that are too high. A similar problem can occur when measuring for trace amounts of sodium from samples stored in soft glass containers. As discussed earlier, due to static electricity, fines from one sample can end up in the next sample if the sampling tool is not cleaned in between.

Atoms or molecules of the component of interest may be added to the sample if a chemical reaction occurs. Physical additions can also occur. If the property of interest is the particle size distribution or a particular size class, material left in a mesh screen from a previous sample will add to the percent weight of that fraction. Other fractions will also be affected as a result.

When the component of interest is moisture content, special precautions must be taken to keep the sample dry so that moisture is not absorbed. Collecting samples in the rain without providing proper protection for them will obviously produce a bias. Gy (1992, pp. 295–296) reports that in a particular uranium plant, samples are handled and split in special climate-controlled environments to avoid changes in moisture content.

Loss

Sample loss occurs when some of the sample mass is not retained after the sample is taken or the percent weight of a particular component is lowered due to chemical alteration. Spills and splashes can happen during or after sampling. Static electricity that is the source of contamination (addition) of fines or flakes from one sample to the next is also the cause of a loss of material when they are not removed from the sampling tool. This results in an especially large bias when we are interested in particle size distribution, including the percent of fines! Mesh screens on the production room floor or in the lab can retain particles or dust from the sample. Precious metals can smear onto crushing or grinding equipment and not remain part of the sample. With certain liquid and gas samples, containers may need to remain cold and the lids secure to prevent evaporation or chemical alteration.

Unstable materials are especially susceptible to reaction with various forms of oxygen and water, reducing the content of certain components. Physical loss can also occur. If we are interested in moisture content, we must make sure that the samples are not left in conditions where evaporation can take place. If the property of interest is the particle size distribution or a particular size class, then reducing the number of transfers from one container to the next will reduce the loss of fines. Smaller particles and fines are easily lost in outdoor sampling with front loaders and shovels. Fines and dust can cling to scoops. Sample lines connecting the sample point to the sample container can also retain smaller particles or dust. Larger particles can be caught in the elbows of sample lines.

Unintentional mistakes or intentional tampering

Sample integrity can be compromised because of operational or analytical measurement errors that are unintentional. Intentional tampering may also be a possibility.

Unintentional mistakes are innocent but can bias the sample results. A sample that is mislabeled as not requiring refrigeration, for example, may be left at room temperature and bias the lab analysis. A sample tag that falls inside a liquid sample is a source of contamination. Samples may be mixed up. Part of a sample may be spilled and not recovered. Sometimes unintentional mistakes can be identified by simply observing the sampling process. We may find that people are unaware of correct principles and procedures. They may use shortcuts, not realizing the negative impact that these actions can have. *Educating* people on the principle of correct sampling and the consequences of not following it is one way to increase the chances that better samples will be taken. Subsequent *training* on proper tools, techniques, and procedures will reinforce the practical aspects of good sampling. These two efforts, education and training, should reduce the occurrence of unintentional mistakes.

Deliberate tampering is fraud or sabotage. Selective sampling can result in biased environmental measurements. Chemical analyses may be falsified to save time if the lab is overloaded with samples. Samples may be purposely put in metal containers to increase the measured metal amount. Intentional tampering can sometimes be identified by a trained eye during a routine audit. In other cases, surreptitious observation may be necessary.

3.8 Summary

The essential idea in this chapter is the *principle of correct sampling*:

- *Every part of the lot has an equal chance of being in the sample and*
- *the integrity of the sample is preserved during and after sampling.*

Always examine tools and their use in light of this principle. Know whether it is violated, how it is violated, and to what extent it is violated. There are several things we can do to improve our chances of following this principle in practice.

1. *Reduce the sampling dimension if possible.* In practice, correct sampling of three-dimensional lots is not possible, so we must try to sample across one or two of the dimensions. Sampling the lot during transfer, before it becomes three-dimensional, is generally preferable.

2. *Define a correct sample.* Every part of the lot must have the same chance of being in the sample. We must use the correct geometry for the sampling dimension: a cylinder for two-dimensional sampling and a slice for one-dimensional sampling.

3. *Condition (mix) a one-dimensional enclosed liquid or gas stream.* Since a cross-stream sample is impossible to obtain, giving the stream a more homogeneous cross section reduces the grouping and segregation error (GSE) and the impact of incorrect delimitation. It also reduces the impact of taking an incorrect sample.

4. *Choose the right sampling tool.* Examine the tool and look for violations of the principle of correct sampling. The tool must be capable of taking the sample that is defined and taking it correctly.

5. *Use the tool correctly.* Just because a tool meets all the theoretical requirements does not mean it will operate properly under adverse conditions or that it will be used correctly by those not properly trained in its operation. Observe the tool in action.

6. *Preserve the integrity of the sample.* The most careful work to get a correct sample using correct tools and techniques is lost if the chemical or physical properties of interest are allowed to change during or after sample collection. A chemist is the best resource to determine how the sample might be susceptible to such changes and how they might be avoided or minimized.

Chapter 4

The Process: Sampling and Process Variation

4.1 Variation over time

Plotting process data is an important and necessary step in understanding process variation. Both simple and complex techniques can be used. In this chapter, we discuss two techniques for analyzing data collected over time. The first is a very simple technique, called a *time plot*, and the second, more elaborate one, is called a *variogram*. More sophisticated statistical time series tools can be used and are often advantageous, and they should be considered if a more theoretical understanding is indicated. Since our intent here is to present basics, we will not discuss any advanced procedures and will give only an introduction to the variogram.

We have seen that two sources of sampling error are the variation of the material as a *short-range* or *localized* phenomenon: the FE, and the GSE.[12] Variations, such as *cycles*, *long-range trends*, and *nonrandom changes*, result from differences in the material *over time*. Changes in the process, either intentional or incidental, result in variation, and samples taken sufficiently far apart in time may differ from each other substantially in the properties of interest. If we do not characterize the process variation relative to the material variation, our ability to understand and control the process or to reduce its variation will be limited and, in some cases, futile.

We examined the fundamentals of one-dimensional sampling in the previous chapter, and we can characterize variation in time as one-dimensional. A one-dimensional *stream* in time can be moving or stationary. It can be a stream of solid particles, a liquid, or a gas. The liquid can have suspended solids and the gas can have suspended liquids or solids. Such material is difficult to sample correctly, even in the controlled environment of the lab. In addition to solid or

[12]Recall that there is also variation in the measurement procedure, though this is not a sampling error.

amorphous matter, we might also have a set of discrete units such as rail cars, bags, or jars. These examples may appear to be different, but in fact they are identical both from theoretical and practical viewpoints. We can identify three sources of variation in these scenarios:

1. short range (mostly random and localized), which includes sampling errors due to the material variation (FE and GSE), sampling errors due to a failure to follow the principle of correct sampling, and errors due to chemical or physical lab analysis;

2. long range (nonrandom), such as trends; and

3. cycles (nonrandom).

These are generally independent and concurrent contributors to the overall variation we observe in measuring the properties of interest. They are thus *additive*. A time plot and a variogram can help identify and quantify these contributors to the process variation.

4.2 Time plots

A time plot is a graph of the measurement of interest plotted against time. It is a simple tool whose power is often overlooked and consequently underused. Shifts, trends, and cycles as well as unusual observations (*outliers*[13]) can be revealed in a time plot. An estimate of the variation can also be obtained.[14]

An example of a time plot is shown in Figure 4.1.

We see that the measurements vary between about 2 and 8, and that there is no obvious pattern. We can gain more information by using different symbols or colors to represent different work shifts, targets, products, etc. Suppose, for instance, that the data in Figure 4.1 were obtained from 8 h work shifts. We can indicate each shift by a different symbol, as shown in Figure 4.2, and see that measurements from shift 1 are consistently higher than the others. An investigation into this phenomenon should be pursued.

If a time plot shows a shift, trend, or cycle, then in addition to examining the process more closely, we should investigate the sampling frequency and sampling techniques. The importance of the effects of sampling in such cases is often overlooked. Several examples follow. In each case, the process itself could be causing the observed pattern in the time plot, but we show how sampling can also be the culprit.

[13] An *outlier* is a value that is somewhat higher or lower than the others. It can result, for example, from an error in chemical analysis or a transcription error. It can also be a real value indicating that something very different occurred in part of the process.

[14] An extension of a time plot is a *control chart*, which can be used to examine the natural variation in the process and to flag unusual situations. Classical control-charting philosophy and usage are well documented in the literature and will not be examined here. Integration of classical control charting with sampling theory can be found in Pitard (1993).

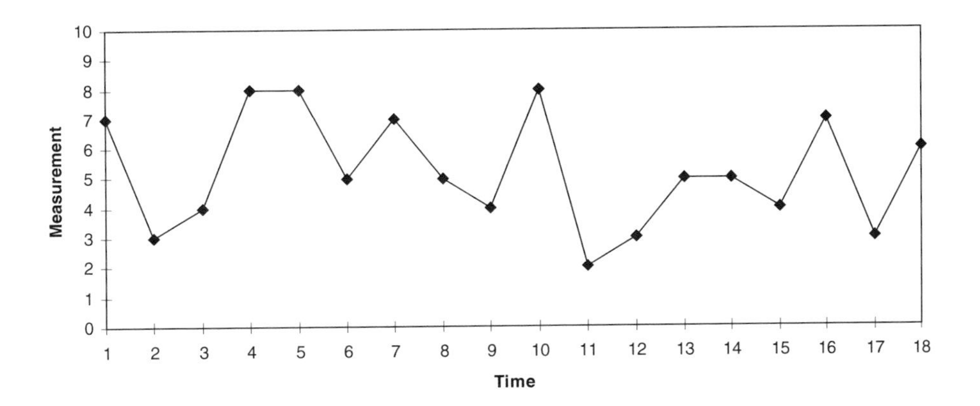

Figure 4.1: Generic time plot.

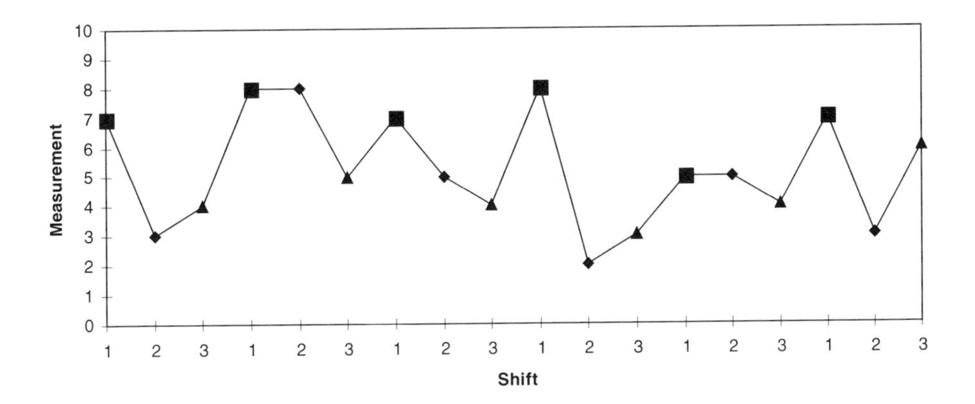

Figure 4.2: Shift 1 reports higher values.

Example 1: Sampling too infrequently

The time plot in Figure 4.3 shows a long cycle.

We can look at process details from all angles and never discover the cause. If, however, as a routine part of the investigation of a cycle, we also examine the sampling frequency, we can discover the cause in this case. By increasing the sampling frequency severalfold over a defined period of time as a check, we see another cycle that is much shorter, as shown in Figure 4.4.

If we overlay these two time plots, we get the graph shown in Figure 4.5.

Now we see that the original long cycle is purely artificial and a result of sampling too infrequently.[15] The process has a cycle, but not the one we originally thought was there.

[15] In time series analysis the false pattern due to sampling less frequently is known as *aliasing*.

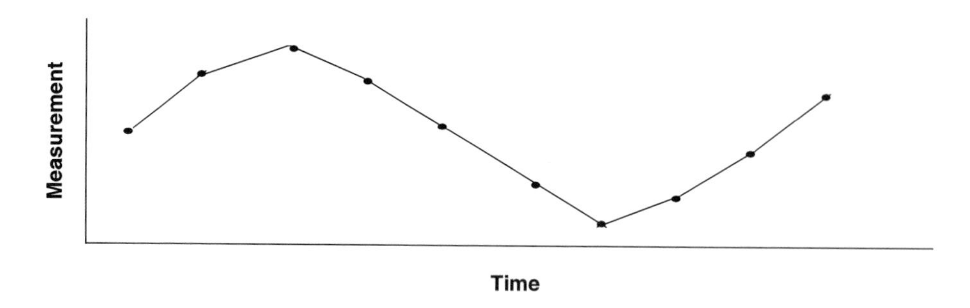

Figure 4.3: Time plot with long-term cycle.

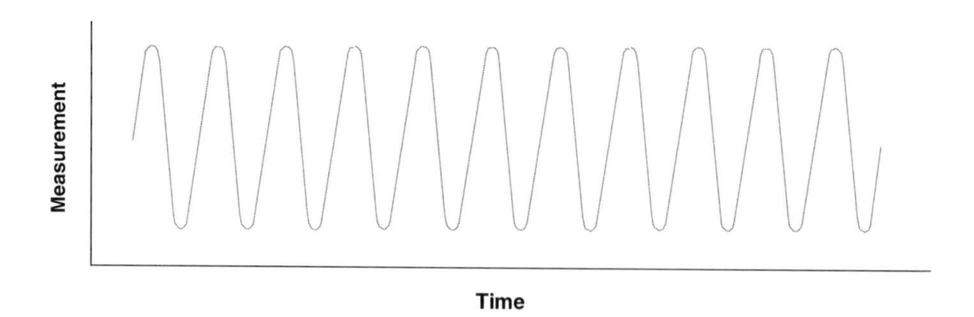

Figure 4.4: Time plot with increased sampling frequency.

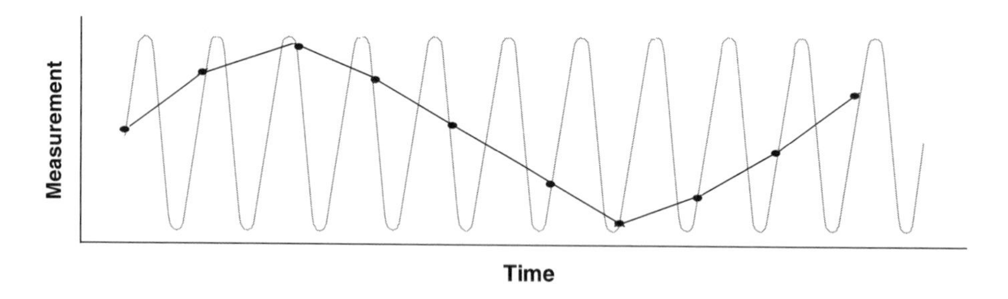

Figure 4.5: Process appearance from two sampling frequencies.

Example 2: Sampling coincident with the same part of a process cycle

Suppose we unknowingly sample with a frequency that is approximately synchronous with a process cycle at low measured values, as shown in Figure 4.6.

Then several problems arise of which we are completely unaware.

1. We will not see that the process cycles and that it cycles with a regular frequency between 1 and 7.

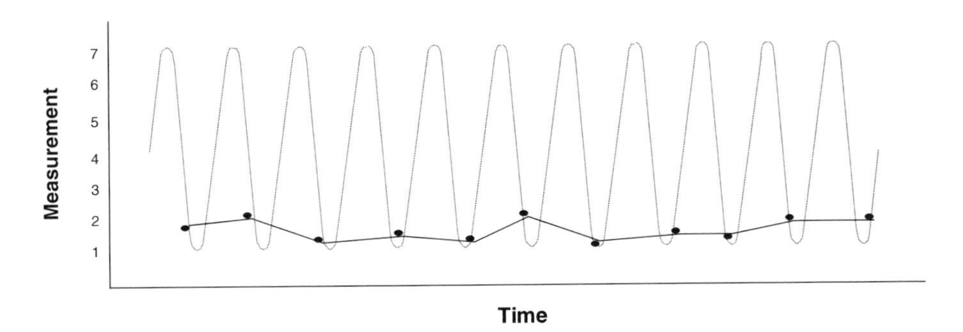

Figure 4.6: Sampling synchronous with a process cycle at its low values.

2. We will not see all the variation in the process—only low values.

3. We will get an optimistic view of the variation, thinking our measurements hover between 1 and 2.

4. We will get a biased estimate of the measurement of interest—only low values.

These situations can cause problems with customers since their sampling techniques may show the product to be off spec. Or, if our process is sending waste to a flare or treater in a cyclical fashion of which we are unaware, then we may think (incorrectly) that we are in compliance with environmental regulations. A spot sample by a local, state, or federal agency may reveal otherwise. A hefty fine might be the result.

Example 3: Sampling that does not follow the principle of correct sampling

Suppose our measurements alternate and cluster around two values: one high and the other low, as illustrated in Figure 4.7.

There is clearly a cycle. The question is whether it is real or artificially induced. In other words, we need to know whether the process itself is actually fluctuating or whether there is something in the sampling that produces these results.

In this case, measurements are derived by taking pellet samples from a conveyor belt once during each 12 h shift. A particle size distribution analysis is performed in the lab. Using a sieving process, the weight percent of "overs" (pellets that are too large) is calculated. The pattern in Figure 4.7 results from (1) segregation by particle size across the belt and (2) day and night shifts taking samples on opposite sides of the belt. Since one side of the belt has a higher weight percent of overs, the samples of one shift will have higher values than the other. This accounts for the alternating sequence of measurements.

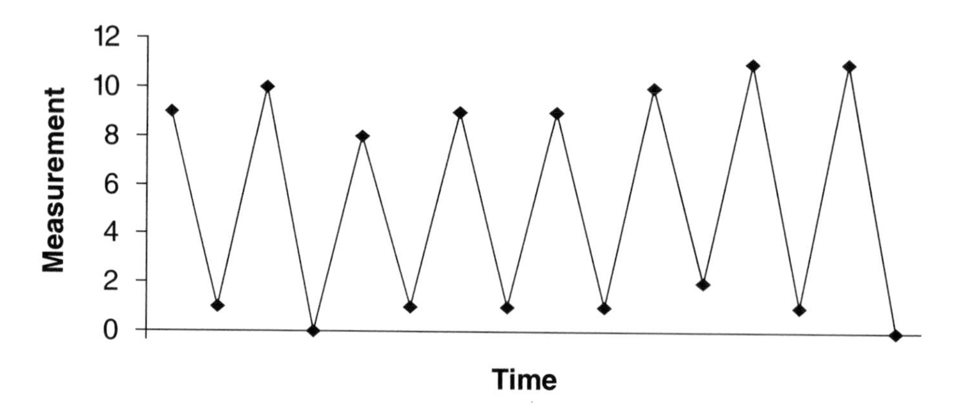

Figure 4.7: Alternating sample results.

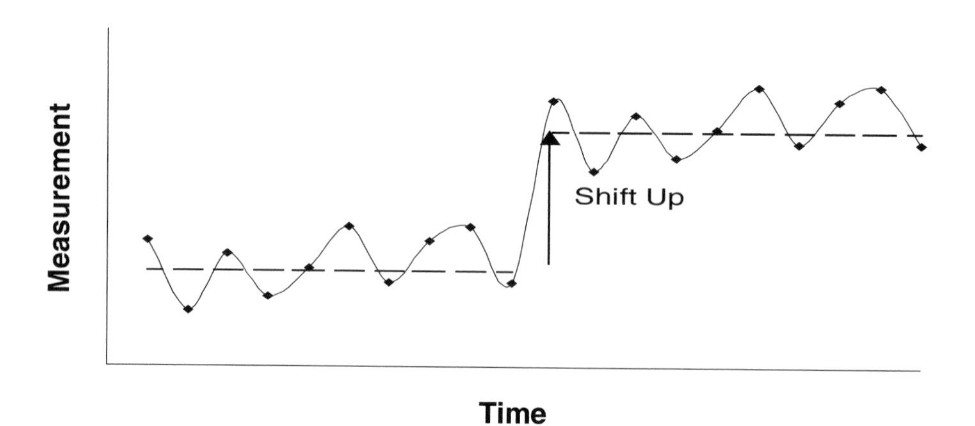

Figure 4.8: Time plot with shift.

Thus, the cycle observed in the time plot is not due to the process; the cycle is due entirely to the sampling. The process itself is very consistent, though perhaps unsatisfactory, in producing pellets that are distributed by particle size across the belt.

Example 4: Improper sample handling

Suppose we observe the time plot in Figure 4.8, showing a shift in the data from low to high values.

A process change could certainly cause this. If we did not do anything intentional in the process, however, then we must look for other, hidden process changes *as well as look for changes in sampling.* It's possible that a restricted sampling line was cleaned out or that a blockage in the sampling line broke free. Or perhaps there was a change in the glassware being used to collect the sample, and the percent weight of the water or sodium content was affected.

4.3 Time plot summary

These examples are used for illustration purposes only, and we do not mean to imply that the explanations given will always fit scenarios you will see in your own processes. Certainly, as stated earlier, process changes could have resulted in the time plots shown. However, the point we wish to make *emphatically* is that sampling should be on any checklist for troubleshooting process problems and should be taken seriously as a possible cause for problems.

4.4 The variogram

Another and more elaborate time series plotting and analysis technique is the variogram. *A variogram is a graph of the variation of samples taken at regular frequencies plotted against the time lag between samples.*[16] We might plot, for instance, the variation V(1) of samples taken 1 h apart (time lag = 1), the variation V(2) of samples taken 2 h apart (time lag = 2), etc. Each point in the variogram is thus the variation of sample results taken a certain number of hours apart. To allow comparison to the variance of the FE (discussed in Chapter 2), which is relative and dimensionless, division of the variogram values by the squared average of all observations is needed. A mathematical formula and explanation are given in Appendix D, including an example using an Excel macro.

Figure 4.9 shows a variogram that indicates a trend in process variation, a cycle in process variation, and a leveling off of the variation. All variograms do not have all these features, but obvious patterns are not difficult to interpret. A variogram can be used to identify the presence of trends and cycles in the variation, assess the periodicity of cycles, and determine proper sampling frequency.

A variogram may reveal data structures that are not apparent from a time plot. Sometimes a time plot provides a lot of information, as we have seen in the previous examples, and a variogram does not add much more. In other cases, the time plot does not reveal trends or cycles, but the variogram shows patterns and thus adds to our understanding of process behavior. The time plot in Figure 4.10 shows process variation with values between 1 and 6 but no regular cycle. The corresponding variogram of the data in Figure 4.11, however, indicates a 5-lag cycle in the variation with a minimum variation every 5 lags.

This means that samples taken 5 time units apart, regardless of the starting point, have more consistent values than samples taken closer together or farther

[16] Variograms can be calculated from irregularly spaced data. However, the plots can be misleading if not interpreted with care, so this case is not addressed in this introductory primer. Variograms can also be plotted with a *distance* lag between samples. Examples include applications in geostatistics and environmental solid waste studies (Cressie, 1993).

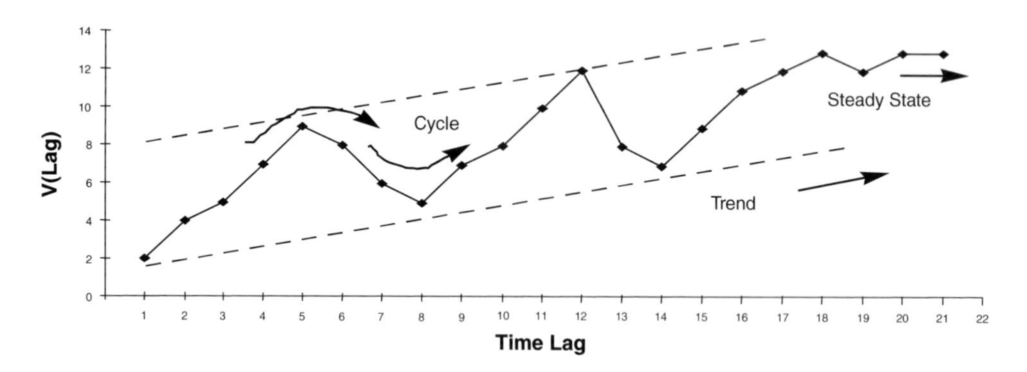

Figure 4.9: Generic variogram.

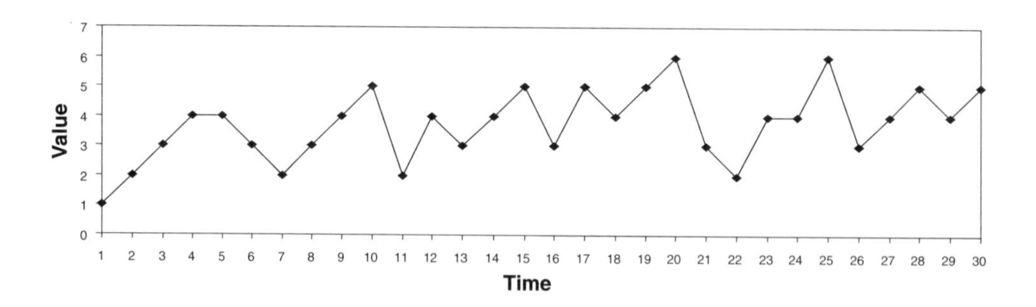

Figure 4.10: Time plot does not show a regular cycle.

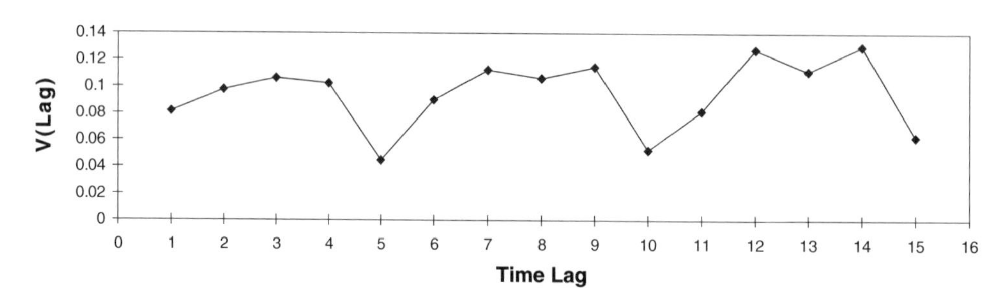

Figure 4.11: Variogram reveals a cycle of five time units.

apart in time. To illustrate this, Figure 4.12 shows values from samples taken 5 time units apart but plotted consecutively. One line has values from times 1, 6, 11, etc. Another has values from times 2, 7, 12, etc. For all starting points except 2, the values are fairly consistent. That is, there is not much variation. In contrast, Figure 4.13 shows values from samples taken 3 time units apart. One line has values from times 1, 4, 7, etc. Another has values from times 2, 5, 8, etc. In this case, the values are not consistent for any starting point.

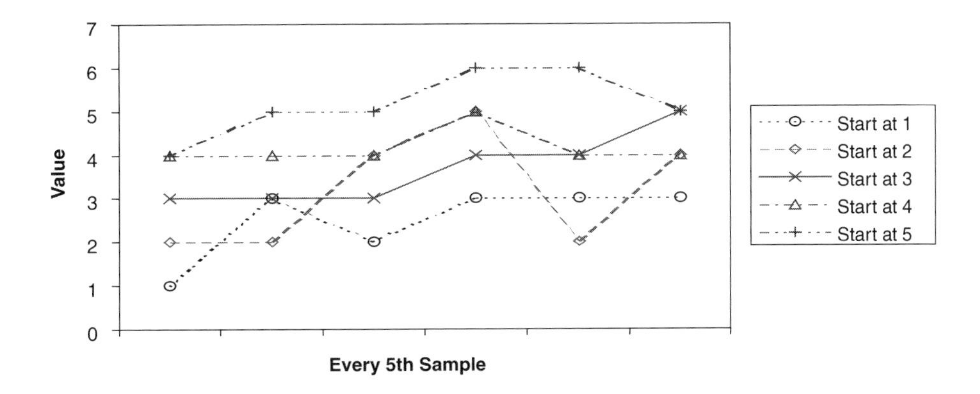

Figure 4.12: Samples five time units apart are consistent for the most part.

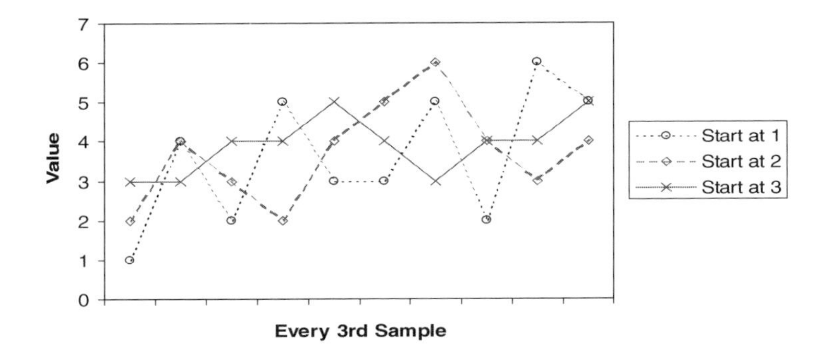

Figure 4.13: Samples three time units apart are not consistent.

4.5 Intuitive explanation of the variogram

We give here a simple example to demonstrate the basic idea. Suppose we are measuring our product for impurities and sample the process hourly. For illustration purposes, we use a circle to represent each sample, and the degree of shading differentiates the impurity level. We observe a pattern in the results of Figure 4.14, so that a variogram is not really needed. However, we use this example to show intuitively how the variogram is calculated and interpreted.

If we look at any two samples taken consecutively, that is, 1 h apart, we see some difference in impurity level, but not much. The difference from one sample to the next one is not great. If we look at samples 2 h apart—samples 1 and 3, samples 2 and 4, etc.—we see larger differences and thus larger variation. Thus, the combined or *average* variation is larger for samples 2 h apart than for

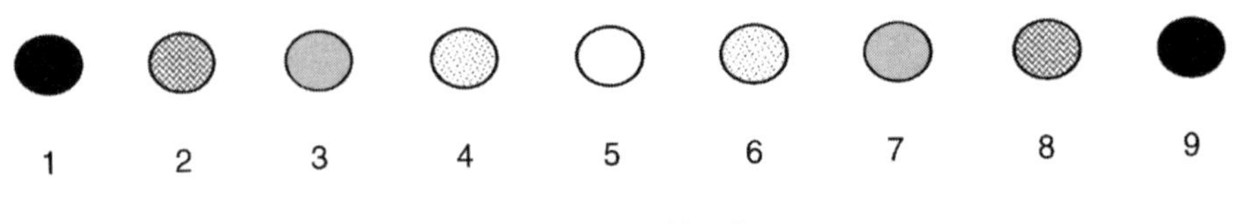

Figure 4.14: Impurity measurements over time.

samples 1 h apart. Intuitively, this means that samples taken closer together in time are more alike than samples taken farther apart in time.

Differences between samples tend to increase up to a point, and then generally one of two things happens.

1. The difference in time between samples coincides approximately with a process cycle.
2. The samples are far enough apart in time that we are just as likely to get a high impurity sample as a low one, or one in between. In other words, the variation in process trends or cycles is averaged out because the values are not correlated, and the variation reaches a steady state in the variation.[17]

Figure 4.9 shown earlier is a variogram that illustrates a trend, a cycle, and a steady state in the variation.

4.6 The nugget effect

Extrapolation of the variogram plot to the vertical axis gives a value called the *nugget effect.*[18] Theoretically, this is the mathematical limit of the variation of differences between lagged samples taken closer and closer together. It contains the sources of sampling variation we have discussed previously: material variation (in Chapter 2, the FE and the GSE, both of which are always present), the variation due to sample collection (in Chapter 3, errors arising from incorrectly defining, extracting, and handling the sample), as well as the nonsampling variation due to chemical or physical analysis.

Estimation of the nugget effect can provide valuable information about the process. For example, if it is substantially larger than an estimate of the material variation obtained independently, then we know that extraneous variation and/or bias is being introduced through incorrect sample collection, handling, or unacceptably large analytical variation. Extrapolation of the variogram to

[17] In geostatistics, this is called a *sill*, characterizing the global heterogeneity of the process (Gy, 1998, p. 92).

[18] The term *nugget effect* is attributed by Cressie (1993) to Matheron. It describes the variation between two closely spaced samples due to the possible presence of *nuggets* of the particular mineral of interest.

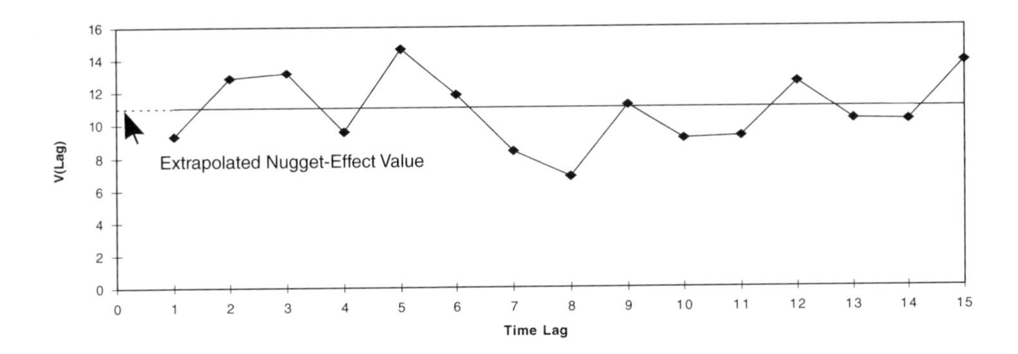

Figure 4.15: Extrapolation of the variogram to estimate the nugget effect.

estimate the nugget effect can be accomplished in different ways, and some techniques are elaborate. Gy (1992) and Pitard (1993) give details and examples on how to do this. Francis Pitard Sampling Consultants (1995) also has software to do this. The easiest method is to take process samples that are only seconds or minutes apart. The variogram points should vary around a horizontal line, which can then be extrapolated as shown in Figure 4.15. This is valid only if we know there is not a cycle in the first few points.

4.7 A caution about irregular sampling or missing samples

While the variogram is defined as a continuous mathematical function of time or space for continuous processes, continuity "is a mathematical concept that does not exist in the physical, material world" (Gy, 1998, p. 87). We must therefore be careful in its estimation. When collecting discrete data and using the formula in Appendix D, the variogram makes sense only if the following two conditions are satisfied.

1. Due to the nature of the formula, the time differences between samples must be constant. That is, samples must be taken at (approximately) equally spaced time intervals. A variogram will be misleading if calculated from samples taken at irregular intervals, such as at noon, 1:00 P.M., and 8:00 P.M. on one day, then 6:00 A.M., 10:00 A.M., and 4:00 P.M. the next day, and continuing in an irregular fashion.

2. Missing samples must be accounted for. In practice, the scheduled sample may not be taken, the chemical analysis may be questionable and thus not used, or the sample may deteriorate because its delivery is delayed so

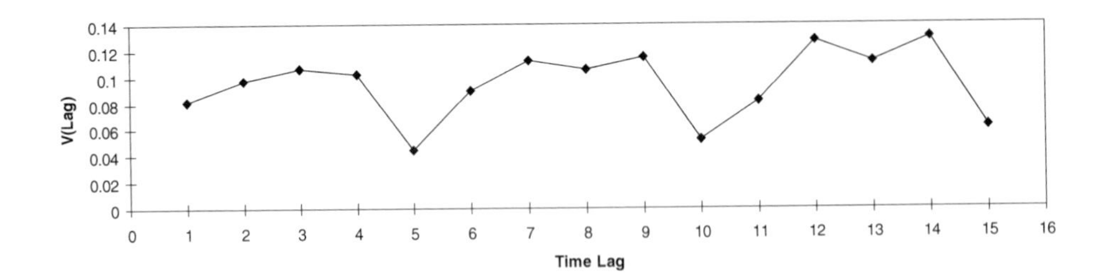

Figure 4.16: A variogram that does not account for missing values shows a cycle five time units long.

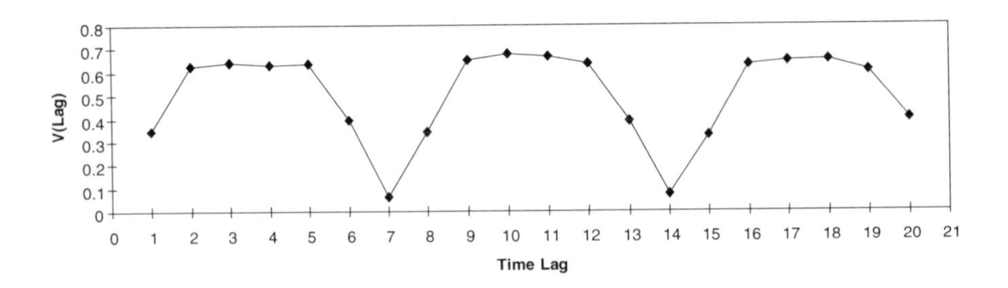

Figure 4.17: A variogram that takes account of missing values shows a cycle seven time units long.

it is not analyzed. The variogram must take these missing samples into account by putting a placeholder in the calculations.

Sometimes samples are skipped intentionally. Suppose, for example, a governmental agency requires weekly samples for environmental monitoring. The plant takes these samples each Monday for reporting purposes. In addition, to monitor the effluent more closely for a few weeks after a major change in the process has been implemented, samples are taken Tuesday–Friday for internal purposes. Weekend samples are not taken since this monitoring is not required by the governmental agency nor deemed crucial for process monitoring. Samples taken on Friday are three days (lags) away from the following Monday, not one lag away as would be used if the missing weekend samples were not accounted for in the calculations.

Figures 4.16 and 4.17 plot variograms computed from the same set of data as that shown in Figure 4.10. In the first case, nonexistent (missing) weekend samples are not accounted for, while in the second case they are. The first variogram suggests a five-day cycle. The second suggests a seven-day cycle.

If only one or two samples are missing and not accounted for in the calculations, the variogram will not be severely affected. However, it is best to make sure that these exceptions are not ignored.

4.8 Examples

Let us examine a few special cases to illustrate how different variograms might be interpreted. Then we will see how the variogram can improve our process understanding.

Example 1: Nugget effect is too large

Suppose we normally take a sample at the beginning of every 8 h work shift. Using historical data, we construct the variogram shown in Figure 4.18.

If the process variation is unacceptably large, we may be tempted to sample more frequently to control the process more tightly. To test this theory, we collect samples every hour for several shifts for a variographic experiment. We also collect a sample every minute for 30 min to be able to estimate the nugget effect (similar to the technique shown in Figure 4.15). We construct the variogram shown in Figure 4.19 based on the hourly samples, using the nugget effect estimated from the minute samples.

We see that about half the variation between samples taken at the usual 8 h intervals is due to the nugget effect. While we will observe less variation between samples by sampling more frequently, most of the variation will be due to the nugget effect, not to the process. And we can never control the process below this nugget-effect amount because it is not due to the process. The variation associated with the nugget effect is due to all the other sampling errors and the analytical error. If we did not take the time to estimate the nugget effect and understand its contribution, we might manipulate the process in various ways in a futile attempt to obtain very tight control. To reduce the nugget effect, we have to examine the material variation, look for problems in sampling tools and techniques (violations of the first part of the principle of correct sampling), sample handling (violations of the second part of the principal of correct sampling), and laboratory analyses.

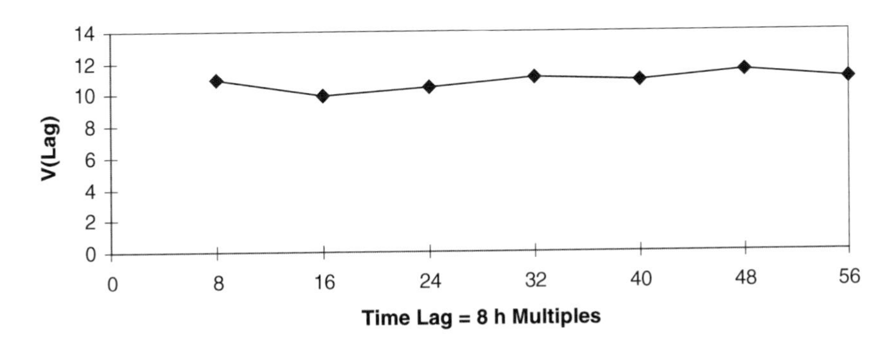

Figure 4.18: Variogram based on sampling every 8 h.

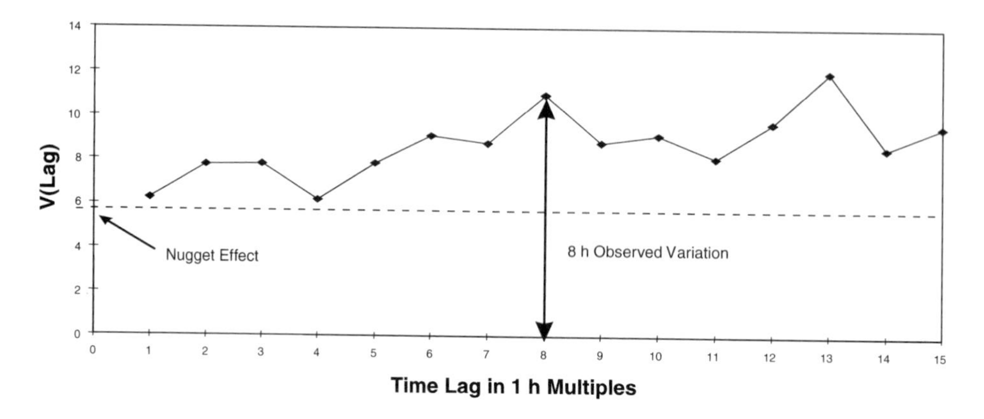

Figure 4.19: Variogram based on hourly samples with nugget effect extrapolated from minute samples.

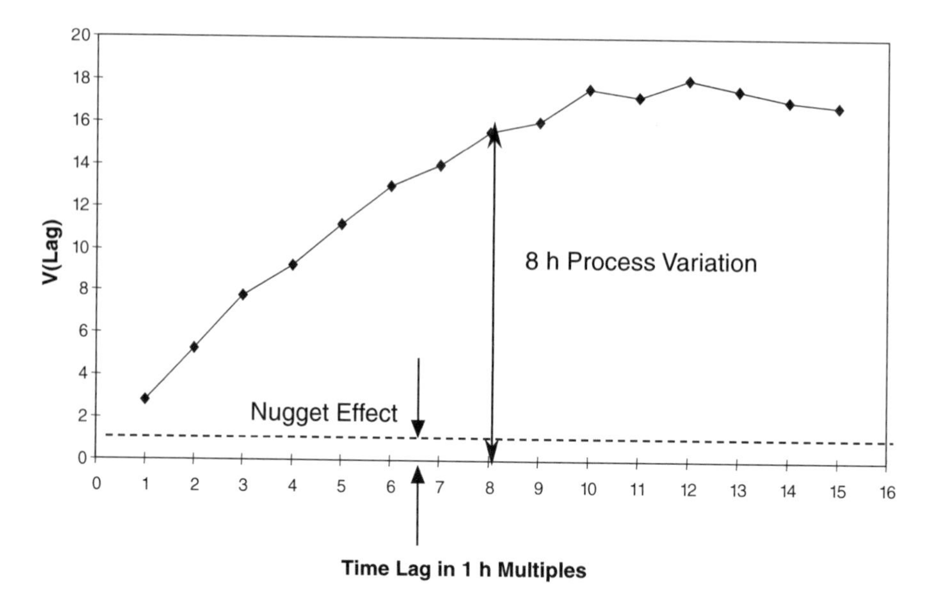

Figure 4.20: Nugget effect is much smaller with a better analytical method.

Suppose we can reduce the analytical variation by more than half by using a more precise method. This reduces the nugget effect substantially, and we perform another variographic experiment. By overlaying all the information, the variogram looks like that in Figure 4.20.

Now most of the variation between the 8 h shift samples is due to the process, not other sources, and more frequent sampling can have a much greater impact on process control.

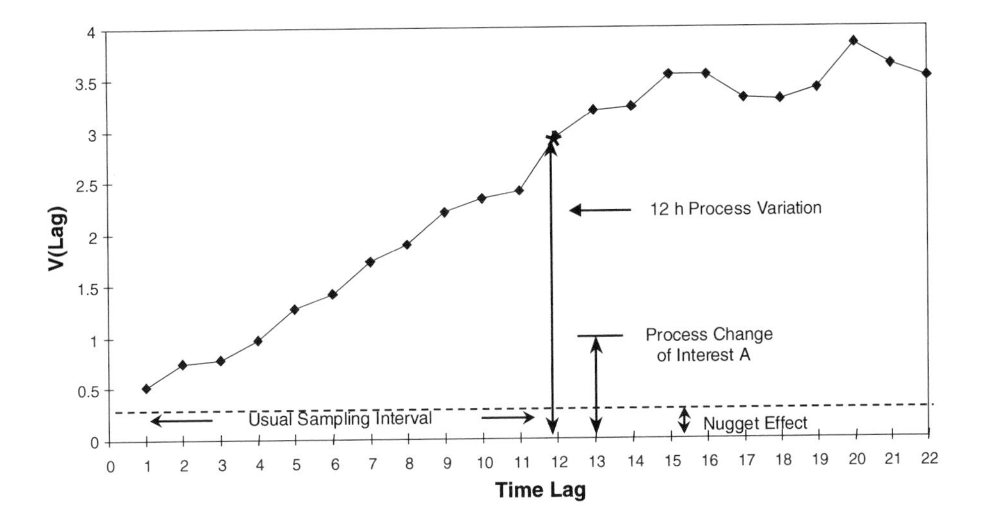

Figure 4.21: Sampling too infrequently to detect desired process change.

Example 2: Sampling too infrequently

Suppose a plant operates on 12 h work shifts and that samples are normally taken at the beginning of each shift. For our variographic experiment, we collect a sample every hour for several days and estimate the nugget effect with a sample every minute. Combining this information results in the variogram given in Figure 4.21.

The total variation between 12 h shift samples, which is what we would normally observe, is indicated by the * and given by V(12). The variation of 1 h lags in between is known only because of the more frequent sampling we performed for the variographic experiment. If we want to control the process variation to amount A as shown, then we cannot do it. The total variation between the usual 12 h samples is too large. However, the nugget-effect value is much smaller than A. So, unlike Example 1, we know it is possible to detect the desired change without reducing the nugget effect. By increasing the sampling frequency to every 4 h, as illustrated in Figure 4.22, we will be able to detect the desired change.

Example 3: Randomness

Figure 4.23 is the variogram for the data from Figure 4.1 over a longer period. There are two ways we can interpret it.

1. *The process variation has no trends or cycles in the time period examined.* We do not see a definitive pattern. It is possible, however, that collecting data over a longer period of time would reveal something. If the total

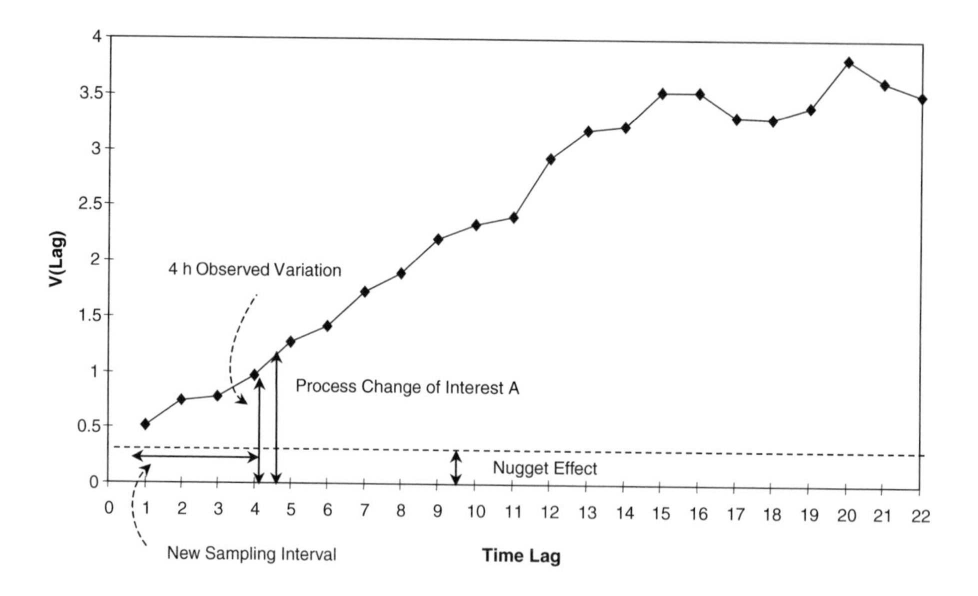

Figure 4.22: Increased sample frequency to detect desired process change.

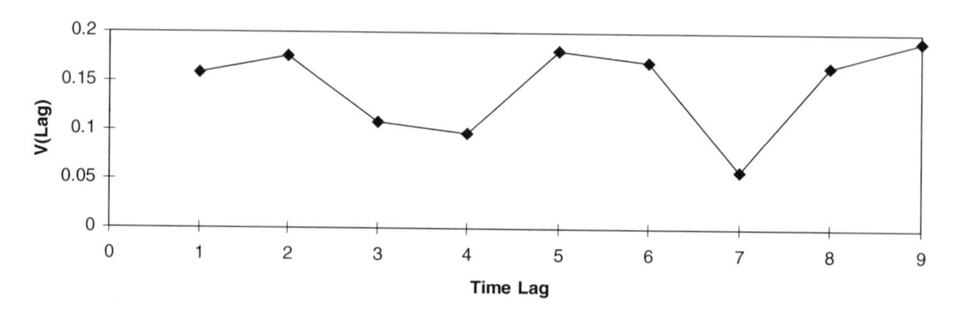

Figure 4.23: Variation appears random.

variation and the nugget-effect value are acceptable, then the sampling frequency is acceptable.

2. *The lags are too far apart to identify trends and cycles in the process variation.* If the overall variation is acceptable, then this is not a problem. If, however, the overall variation is unacceptably high based on customer specifications or other factors, then a variographic experiment should be conducted using increased sampling frequency.

Example 4: Trend

The variogram in Figure 4.24 shows a trend in the process variation.

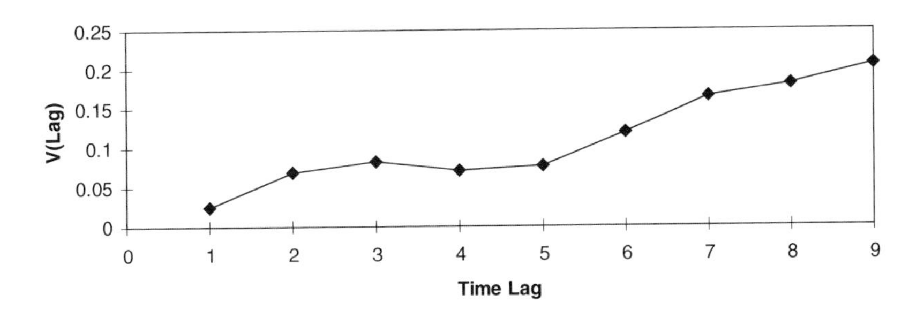

Figure 4.24: Trend in process variation indicated.

The possible interpretations are similar to those for Figure 4.23, discussed above.

1. *The process variation has no cycles in the time period examined.*
2. *The lags are too far apart to identify cycles in the process variation.*
3. *The process has a trend.*

4.9 Use of historical data for time plots and variograms

A time plot and a variogram of historical data provide a good first step in understanding process variation. These may reveal trends or cycles that were previously unknown. If the process has been unacceptable, these plots can provide directions for improvement.

One very good source of historical data for liquid or gas continuous processes is on-line process data that are recorded every minute or perhaps more frequently. Because the data are equally spaced, they lend themselves naturally to variographic analysis. Data from solids processes, batch processes, parts manufacturing, or lab analyses may not be as frequent, but as long as the samples are taken on a fixed schedule, then a variogram can be constructed. A time plot can always be constructed regardless of the sample schedule, though cycles in the process or in its variation may be difficult to determine if the sampling is irregular.

In some cases, historical data may not be appropriate for computing a variogram, or if so, may not reveal any useful information. Below are some possible reasons.

1. Samples are taken so infrequently that relevant trends and cycles are hidden.
2. Samples are not taken and handled correctly, so inflated variation is observed.

3. Samples are taken at irregular intervals, so calculating a variogram using the formula in Appendix D due to lagged differences is meaningless.

4. The process has changes in level like that shown in Figure 4.8.

Pitard (1993, Chapter 26) combines aspects of the time plot and variogram to produce a new kind of chart in a subject area Gy (1992, p. 104) calls *chronostatistics.* It resembles a control chart but is not used to track results in real time. Rather, it is applied to past data to analyze the behavior of the process and used by managers as a variation investigation tool.

We have covered here only the basics. Theoretical details as well as practical applications can be found in Gy (1992) and Pitard (1993).

4.10 Summary

1. *Plot the data against time.* Plot measurements as a function of time on a time plot or expand it into a control chart. Look for nonrandom patterns, sequences, and unusual values. Search for causes.

2. *Plot the data variation against lags in samples using a variogram.* Identify nonrandom changes and cycles. Examine the effect of sampling frequency. Separate the process variation from other sources of sampling variation and also from the analytical variation.

3. *Plot the data in whatever ways are useful to gain insight.* Plotting two sets of measurements against each other in a scatter plot can help identify correlations. Plotting different measurements on the same time line can help find causes for unusual events by the alignment of certain results.

Chapter 5

A Strategy: Putting Theory into Practice

5.1 A logical approach

In previous chapters we detailed correct sampling principles, analyses, tools, and techniques. We also illustrated traps and pitfalls that result in poor sampling. We did this by presenting a structure for the different sampling errors, which is useful for assessing particular situations. We now know that sampling is made up of components and that these components contribute to the overall sampling error. We have seen ways to improve many sampling situations and reduce the sampling error.

Now we need to understand how to apply these ideas as a *strategy*, to go beyond particular problems that are brought to our attention, to address all the components of sampling. In this chapter, we propose an approach to sampling that uses the previous theoretical information in a practical way. The strategy has three parts: *audit*, *assessment*, and *action*.

5.2 Audit

When we learn of a sampling problem, we often focus on an obvious culprit. The particular issue may be a procedure, a collection device, or handling, for example. With this approach, we do not look beyond what we know to be wrong. We are content to solve the "visible" problem at hand. Since we don't see the whole picture, we are oblivious to other sampling issues that may arise in the future. We thus recommend that the first step be a sampling *audit*, allowing us to look at all the sampling components. An audit provides a firm basis for a detailed assessment and, if necessary, action.

Auditing techniques vary somewhat, but we discuss here three useful steps that can be used as a guide.

1. *Examine the written sampling procedures* to find out how the sampling is supposed to be performed.

2. *Walk through the sampling areas* to observe how samples are taken and to ask questions.

3. *Provide both an oral and a written report of the findings* to the sponsors (those who commissioned the audit), to those who have responsibility for taking action, and to those who take samples.

Examine the written sampling procedures

A written procedure should exist for every sample that is taken. There is already a deficiency if any procedures are missing. Every procedure should state

- what the sample is and how much should be taken;
- when, where, how, and why the sample should be taken;
- who should take the sample; and
- precautions for taking the sample safely and with minimal environmental impact.

In describing how a sample is taken, the procedure should give details on tools, techniques, and handling. It should also document frequency and purpose.

Sampling procedures provide an overview of the type of sampling that is conducted. They need to state, for instance, whether the samples are taken for process control, for product release, or for environmental reasons. The procedures alert us to what particulars to look for in the field when we do the walk-through. We can ask those who sample how they were trained to sample and if they are familiar with the procedures. We can compare what the procedures state to what we observe and then note any discrepancies. It is not uncommon to find differences between what is written and what is practiced. These differences may have important economic, safety, or environmental consequences.

Written sampling procedures also provide important documentation for training. Without them, training in how to take samples becomes haphazard. Poor sampling follows, resulting in problems in meeting specifications, running the processes efficiently, or operating safely.

Walk through the sampling areas

Every sample has a history, and two key questions are where it came from and how it was obtained and preserved. There is no substitute for first-hand observation and interviews, so we suggest a *walk-through* of areas where samples are taken. Arrangements should be made to observe people taking samples as they normally would. It takes a trained eye, however, to know if anything is wrong. Consequently, familiarity with the written procedures, with the principle of correct sampling and the sampling errors, and with the industrial process or environmental conditions are crucial for the walk-through to be useful. We have already seen the importance of reviewing the procedures. For the walk-through, a check sheet can be very useful as a memory jogger for finding sampling errors. We can get topics for the check sheet from the summaries at the end of Chapters 2, 3, and 4. Subtopics can be added based on the particular sampling involved. Wherever possible, we should look for violations of the principle of correct sampling and for contributions to the sampling errors relating to the material, to the tools and techniques, and to the process as discussed in earlier chapters. See Figure 5.1 below for an example.

In addition to observation, asking questions of those who take the samples is crucial. For example, ask about their training for taking the samples, how they know where to take the sample, how to take the sample, special safety or handling precautions in taking the sample, and what to do with the sample. Also, ask if they know how the sample is used and its relative importance.

Auditor ______________________ Date __________

Sample location ________________ Who took the sample ______________

Description of sample ____

Yes	No	Not Applicable	
			Sample port labeled
			Instructions for taking sample
			Line flushed
			Proper orientation of sample probe for liquids or gases
			Different particle types or sizes
			Mixing
			Clean sampling container
			Compositing
			Contamination
			Safety

Additional description or comments: Material, tools, techniques, handling

Sketch:

Figure 5.1: Example of an audit check sheet.

In addition to knowing what to look for from a sampling perspective, it is desirable to know what to look for from a process and chemical standpoint. This information affects how the samples are taken as well as safety considerations. We do not sample in a vacuum, but for a reason. In an industrial setting, for example, we are trying to find out something about the process. Knowing what chemical components in the sample are being analyzed can provide crucial information about how the sample should be handled. Thus, we strongly suggest not conducting a walk-through alone: *a process engineer and a process chemist should be on the audit team.*

Provide both an oral and a written report of the findings

An oral report should be given to the sponsors and those responsible for taking action within a day or two of the audit. It should highlight key findings. This report provides a quick and immediate overview as well as a sense of how big or small the problems are. While a written report can be shoved in a drawer unread, an oral, face-to-face communication guarantees that the audit results are heard. In some cases, immediate actions are appropriate, such as changing a collection container for safety reasons or because the material it is composed of reacts chemically with the sample contents. Since those who take samples may not read the written report, an oral report would be their avenue for receiving the information. The timing of the oral report is not as crucial, but they need to get feedback on current sampling techniques so that they are not surprised when later asked for their input or when changes are implemented. After all, those who collect the samples have the greatest impact on sustaining any changes that are initiated.

A written report of the audit findings serves several purposes. First, it forces us to think about what we have observed and to organize our thoughts. It puts the whole audit in perspective, allowing priorities for further action to be set more easily. Second, a written report allows wide distribution of the findings. It is useful for managers who need to act on the deficiencies. It provides feedback to those who were interviewed as well as to all those who take samples. People in departments not involved in the audit may see commonalities and thus make improvements on their own. Finally, written documentation provides a basis for action. It is a record that can be referred to for measuring changes and improvements.

5.3 Assessment

From our observations in the audit and our knowledge of the seven sampling errors, we can begin to evaluate the effectiveness of the sampling in two ways.

1. *Analyze the audit results and available data.*

2. *Conduct experiments, if appropriate, to fill in details for more complex circumstances or processes.*

Analyze the audit results and available data

It is very important to understand and address the causes of problems in sampling before jumping to action. Otherwise, the fixes will be superficial and temporary. Problems pointing to deficiencies in procedures or training are especially susceptible to this pitfall. For example, a procedure may be followed that does not provide appropriate or sufficient detail. Or a sampling procedure might not exist, so people sample according to how they were taught, generally informally. The inability to collect samples according to a preassigned schedule may result from task interference, such as the need to perform higher-priority tasks in the same time frame. Simply rewriting a procedure will not ensure that sampling improves.

Available data should also be examined. For example, historic data are often available and can be used to evaluate process variation for trends and cycles using variographic techniques. We can examine control charts in the lab to measure analytical variation and compare it to process and sampling variation. We can compare our release data to acceptance data from customers, if available.

Conduct experiments, if appropriate, to fill in details for more complex circumstances or processes

If a study of available data does not address all the issues raised in the audit, then a second step, experimentation, is needed. The sampling frequency or degree of compositing, for example, may be questionable. By performing a variographic experiment as described in Chapter 4, the effects of more or less frequent sampling can be evaluated. Sample times coinciding with cycles in the variation can be changed. Estimation of the nugget effect can reveal whether the combination of the other sampling errors is unacceptably large. Analyzing the increments that go into a composite as well as taking more increments will provide a measure of variation that will confirm the effectiveness of current sampling practice or indicate how improvements can be made. Observing how customers sample may point to sources of discrepancies in results. Lab experiments can be performed to assess the effects of subsampling or mixing.

5.4 Action

The initial steps will require acting on recommendations in the audit that are straightforward. For example, efforts to ensure that the sampling points are labeled properly and that precautions for safety are present can be started immediately. On the other hand, most improvements are generally not straightforward or easy. People taking the samples need to understand the changes, and those responsible for making the changes will need to provide some training. Changes can affect a very broad area, encompassing various processes, equipment, and employees working different shifts.

If we have performed some experimentation, then we may need to make more involved changes after the results are known. Sometimes we must install new, correct, sampling equipment at considerable cost. Sufficient economic incentives should be present, of course, to recoup these costs. If current practice differs from what is desired, procedures may need to be written or modified.

The action with the biggest impact and that is at the same time the most difficult to implement is a change in the sampling routine that ensures sampling is done correctly. For example, having people clean the sampling equipment and making sure that clean receptacles are provided and used require changes in the way people do their jobs. Changing the sampling frequency in the field will have an impact on the lab routine and personnel. Whatever changes are made, some planned follow-up is essential to ensure that the changes are lasting.

5.5 Summary

The following steps can be followed as a general strategy.

1. *Audit the sampling procedures and practices through personal observations.* Discuss findings with the sponsor and those involved in sampling. Provide a written report.

2. *Assess the findings of the audit.* Use observations from the audit as well as any available data. Conduct experiments, if necessary, to answer important questions raised by the audit or analysis.

3. *Take actions to address problems uncovered by the audit and assessment.* Remember that implementation of recommended actions takes time and follow-up.

Appendix A

Introduction to Gy's Seven Sampling Errors

A.1 A structure for sampling

The first steps in any sampling investigation are audit and assessment: find out what is going on and whether the current sampling variation is acceptable. If not, then some way must be found to reduce it. This would be easier if the total variation could be broken down and the component parts addressed separately. Pierre Gy's theory does this. Gy (1992) decomposes the total variation into seven major components (sources). He calls them errors because sampling is an error-generating process, and these errors contribute to the nonrepresentativeness of the sample. The seven errors are as follows:

1. fundamental error (FE)
2. grouping and segregation error (GSE)
3. long-range nonperiodic heterogeneity fluctuation error (shifts and trends)
4. long-range periodic heterogeneity fluctuation error (cycles)
5. delimitation error (DE)
6. extraction error (EE)
7. preparation error (handling) (PE)

These errors are introduced below, with descriptors matching those in the rest of this book.

Error 1: Fundamental error (FE) (material variation)

The word homogeneous is defined in *Merriam–Webster's Collegiate Dictionary, Tenth Edition*,[‡] as "consisting of uniform structure or composition throughout." To appreciate the variation that poor sampling can generate, we must realize and accept the fact that no material is homogeneous. *Everything is heterogeneous*, even if only at the molecular level. The constitution (or makeup) of the material causes it to be heterogeneous. Gy calls this the constitution heterogeneity (CH). It represents the differences between particles or molecules. The CH of solids is influenced by particle size, shape, density, chemical composition, and other physical properties. Liquids and gases are heterogeneous at the molecular level for these same reasons. Suspended solids will increase the heterogeneity of liquids, and suspended solids and liquids will increase the heterogeneity of gases.

Because of CH, a sample will almost certainly not be truly representative of the whole. Consequently, due simply to the heterogeneity of the material being sampled, a sampling error will occur. This is the fundamental error (FE), from which no escape is possible. It is generally large for solids and negligible for liquids and gases. For solids, Gy has developed formulas to calculate the variance of the FE. These formulas can be used to determine sample mass in different situations.

Error 2: Grouping and segregation error (GSE) (material variation)

Another source of heterogeneity, and thus another source of sampling error, is due to the differences from one group of particles to another or from one part of the lot to another. Gy calls this the distribution heterogeneity (DH). It is caused by the combination of the CH, the spatial distribution of the constituents, and the shape of the lot. Many solids, liquids, and gases are known to settle or stratify because of different properties of their components. Thus, sampling from the bottom versus from the top of a container can generate very different samples because of *segregation*, giving rise to the *segregation error*.

For a fixed sample size (weight), random sampling produces estimates with smaller variance if the particles are taken one at a time (Elder et al., 1980). But this is rarely the case. Instead, samples generally consist of scoops of many particles in the form of powders, grains, pellets, etc. The extra variation introduced by sampling *groups* of particles gives rise to the *grouping error*. The sampling error resulting from grouping and segregation can be reduced by taking many small increments and compositing them to form the sample.

Error 3: Long-range nonperiodic heterogeneity fluctuation error (process variation)

Processes often change over time, sometimes in short intervals and sometimes over a longer time span. This variation can be broken down roughly into ran-

[‡]By permission. From *Merriam–Webster's Collegiate ® Dictionary, Tenth Edition* © 2000 by Merriam–Webster, Incorporated.

dom, nonrandom, and cyclic variation. The random variation is due to the two errors previously discussed: the FE and the GSE. These are often inflated by measurement error and errors 5, 6, and 7 discussed later. Nonrandom variation is due to shifts or trends in the process. Some of this variation may be under control and known, such as a change in the feed to a process unit. Other nonrandom variation may be out of our control and may or may not be known. For example, a severe temperature shift in a processing unit may be observed, and it may take some time to find that the cause is a malfunctioning heat exchanger.

Because of this long-range fluctuation error, samples taken at different times will give different results. Consequently, it is important to determine whether such trends exist and how they behave. Then we can be sure that process adjustments are effective by using appropriate sampling frequencies.

Error 4: Long-range periodic heterogeneity fluctuation error (process variation)

In addition to the random variation and long-range fluctuation errors, processes often experience changes over time that are periodic. For example, the process may be affected by day and night temperature cycles. Batch operations that replace solutions every two runs will often experience alternating sequences of results. This *periodic fluctuation error* affects the variation in the process. Consequently, the timing of the sampling affects how the process is run. The cause of the process cycle is not a sampling error, but a sampling error may be generated by variations in the cycle period, amplitude, and sampling frequency.

Systematic sampling that has the same frequency as the cycles will not reveal the entire variation of the process. It also tends to produce biased results because the values are generally too high or too low, depending on when samples are taken relative to the cycle. Some cycles can be spotted easily in a time plot of the data; others may be more subtle. Some cycles may be very regular, others somewhat irregular. It is therefore important to understand the size and effect of this component. A time series analysis tool that can help identify long-range and periodic errors is the variogram (Chapter 4). A generic example is shown in Figure A.1.

Error 5: Delimitation error (DE) (including the principle of correct sampling)

Nonrandom samples, such as judgmental and spot samples, are useful in some situations—for example, in the preliminary assessment of a toxic waste site.[19] Probabilistic and random samples, however, are fundamental to obtaining *unbiased* estimates if the true average value is being sought. A very basic idea, yet one that is often overlooked, is that to get a random sample, every part of the lot to be characterized must have an equal chance of being in the sample.

---

[19]However, it is important to understand that nonrandom samples can confirm the presence of a substance but not its absence. Also, while they can confirm presence, they cannot confirm (or estimate statistically) specific quantities.

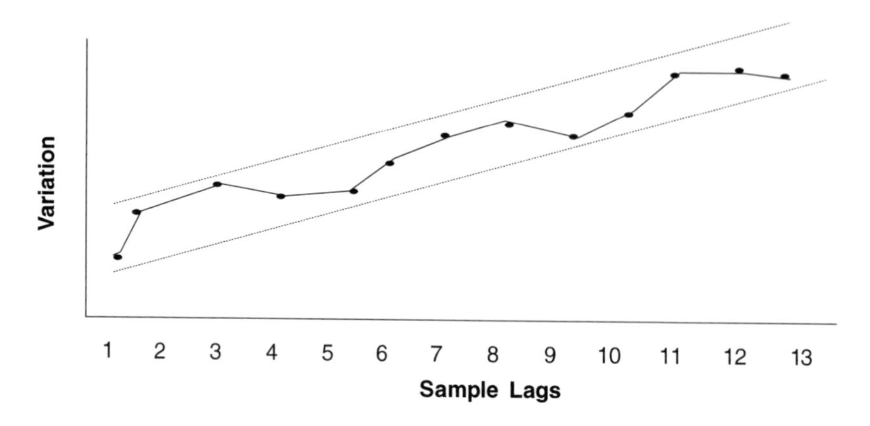

Figure A.1: Example variogram.

A sample defined in this way is called *correct*. A *delimitation error* occurs when not every part of the lot has an equal chance of being in the sample, in other words, when the defined sample boundary is not correct.

This relatively simple notion of equal probability of access has some important consequences. Scooping off the top of a solid pile, for example, or pouring liquid from a container, violates this fundamental principle and may produce biased samples. Results from these types of samples will be especially misleading if the material has a large DH.

Error 6: Extraction error (EE) (including the principle of correct sampling)

Once a correct sample has been defined, it must be correctly taken. An *extraction error* occurs if the sample that has been identified cannot be obtained. In other words, a delimitation error may be avoided by defining a correct boundary for the sample, but if it cannot actually be recovered, then an extraction error is incurred. One of the main impediments to extracting the defined sample is the equipment used. It is especially difficult to sample material in three dimensions such as that in a tank, rail car, landfill, or 100 g of powder in the lab. To obtain a random sample, the bottom of a container as well as the top must be accessible. While sampling from the top is generally easy and from the bottom difficult, getting a theoretically predefined sample of material from the middle is virtually impossible! For liquids, dropping a bottle to the desired depth of a large tank or drum is one way to approximate this, but for solids no practical technique that is also theoretically sound has been developed.

A vertical core sample is a correct delimitation of a solid flat pile if it represents the whole thickness. Using a thief probe to obtain it, however, produces an extraction error because the thief cannot extract material at the very bottom (Figure A.2).

Figure A.2: Thief probe.

If a soil sample is taken with a coring device whose diameter is too small to obtain the largest chunks, then an extraction error occurs.

Delimitation and extraction errors contribute to both bias and variation, and bias is very difficult to detect unless a special effort is made. As a result, the magnitude of these errors is often unknown and frequently underestimated. It would be extremely unfortunate to learn of a bias via a lawsuit. To avoid bias, a proper tool must be chosen and used correctly. Even if the right equipment is available, additional error is added to the total sampling error if the equipment is not used correctly.

Error 7: Preparation error (PE) (including the principle of correct sampling)

Gy uses the term *preparation error*, but we should not confuse it with sample preparation in the laboratory. It is much more general than that and might be better thought of as *sample handling*, *sample integrity*, or *sample preservation*. We know that samples can change as they are obtained and also between the time they are taken and the time they are analyzed. To reduce bias, the sample integrity must be preserved. For example, some samples must be kept cool; others require containers made of special material. Evaporation, alteration, loss of fines, etc., must be avoided to ensure that the sample maintains its integrity. Otherwise, all the effort to characterize, reduce, and avoid the other sampling errors will be lost.

A.2 Grouping used in this primer

For presentation purposes, we have grouped the errors into three general categories and used intuitive descriptions:

- the material (short-range variation), combining errors 1 and 2;
- the tools and techniques, including handling, combining errors 5–7; and
- the process (long-range and periodic variation), combining errors 3 and 4.

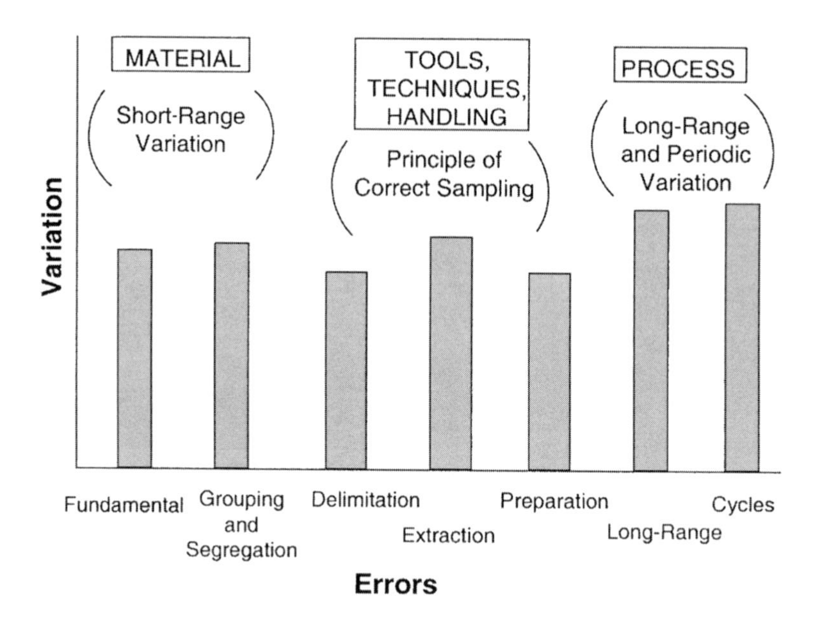

Figure A.3: Example bar graph of variation vs. errors for sampling. (Actual variation depends on situation.)

In this primer, errors 1 and 2 are discussed in Chapter 2, then errors 5–7 in Chapter 3, and last, errors 3 and 4 in Chapter 4. These components are shown in Figure A.3 in a bar graph. In this illustration, the variation is roughly evenly split among the seven major sources. It will be different in other sampling situations. Incorrect sample handling could be the dominant source of variation when sampling a volatile liquid or solid, while material segregation might be the major source when sampling a mixture of immiscible liquids.

Appendix B

The Variance of the Fundamental Error

B.1 Importance

The variance of the fundamental error (FE) is most important in solids sampling. If we could repeat the sampling protocol, it is the smallest sampling-to-sampling variation we can have with a given particle size and sample mass, assuming correct sample definition, extraction, and handling, as discussed in Chapter 3. Its estimation is therefore important to determine the best we can expect with the current sampling protocol. If this variance is unacceptably large, then either a larger sample mass must be taken or else the particle size must be reduced (unless particle size is of interest). Particle size reduction is discussed later in this appendix. Because the mathematics is very complicated, we give here only a sketch of the derivation and how the formula can be approximated. Details are given in Gy (1992) and Pitard (1993).

B.2 Heterogeneity of an individual unit

Gy (1992, p. 57) defines the heterogeneity carried by an individual particle as a function of the relative difference between the chemical or physical measurement of the particle and what the true lot value is. This difference is mathematically weighted by the mass of the unit. In (B.1) below, c_i is the *content* of a given particle, that is, the measurement of the characteristic of interest, such as chemical concentration. It is determined by dividing the mass of the component of interest in the particle by the mass of the particle itself. Using M_i for the mass of the particle, the heterogeneity h_i carried by the individual particle is defined as

$$h_i = \left(\frac{c_i - c_L}{c_L}\right)\left(\frac{M_i}{M_L/N_L}\right). \tag{B.1}$$

Here, the subscript L means lot, and N_L is the total number of particles in the lot. In practice, h_i cannot actually be calculated since we do not generally know the mass M_L of the lot nor the total number N_L of particles in the lot. For the moment, however, let's look at h_i intuitively because this relatively simple formula contains much information.

Note that h_i is larger when the content c_i is farther away from (higher or lower than) the true content c_L in the lot. So when we say the heterogeneity is large, we quantify this with a large value of h_i. Also, the heterogeneity carried by a particle depends on its mass. Thus, larger particles will carry a higher heterogeneity h_i than smaller particles, even if they have the same content c_i. Note that h_i is relative and dimensionless because of the division by c_L and M_L.

B.3 Theoretical formula for the variance of the fundamental error

Suppose we take a sample that is fairly representative of the entire lot. Then the average content c_S in the sample should be approximately equal to the true lot content c_L. If we could repeat the sampling protocol and look at all possible samples of the same weight, then the heterogeneities carried by the particles in the sample would be approximately equal to the heterogeneities carried by all the particles in the entire lot.[20] (Of course, we cannot actually do this calculation because we do not have information on the lot, and we only perform the sampling protocol once.) Because the average of all the lot h_i values is (approximately) zero, their variance is given by $\sum_i (h_i^2/N_L)$. We quantify the constitution heterogeneity (CH) as this variance (Gy, 1992, p. 58):

$$\text{(B.2)} \quad \text{CH} := \text{Var}(h_i) = \sum_i (h_i^2/N_L) = N_L \sum_i [(c_i - c_L)/c_L)]^2 [M_i/M_L]^2.$$

Thus, the more the particles differ from each other in the chemical or physical property of interest, the larger the arithmetic value of the CH. In other words, a large *intuitive* CH is reflected in a large *calculated* CH.

Recall from (2.1) that

$$\text{FE} = (c_S - c_L)/c_L.$$

Here, the lot content c_L is a fixed quantity, but the sample content c_S may vary depending on the particular sample obtained. Further, c_S is determined by the ratio of two quantities that also vary:

$$c_S = (\text{mass of component of interest in the sample})/(\text{mass of the sample}).$$

Thus, examining the variance of the FE is equivalent to examining the variance of the ratio of two statistical random variables.

[20] The word *unbiased* is the statistical term used to describe this property.

By using the multinomial distribution, mathematical approximations, and simplifying assumptions, Gy (1992, p. 362) has derived an approximate formula for the variance of the FE (Var(FE)):

$$\text{(B.3)} \qquad \text{Var(FE)} \approx \frac{(1-P)(\text{CH})}{PN_L} \approx \left(\frac{1}{M_S} - \frac{1}{M_L}\right) \sum_i \frac{(c_i - c_L)^2 M_i^2}{c_L^2 M_L},$$

where a simplifying assumption is that $P = M_S/M_L$ is a constant probability of selecting a group of particles in the sample. From a statistical perspective, if all particles have equal mass, then $P = n/N_L$, where n is the number of particles in the sample and N_L is the number of particles in the lot. If only one particle is selected, then $P = 1/N_L$. Here we see that the sample mass M_S is inversely related to the Var(FE), indicating that a larger sample mass would decrease our sampling-to-sampling variation if we could repeat the sampling protocol. There are several ways to estimate the quantities in this formula for practical applications, depending on the situation. We give below two approximations. Statistical properties of these approximations are unknown.

B.4 General approximation

Experimental results in many situations show a high correlation between particle density and the chemical or physical property of interest. They also show a relatively low correlation between particle size and the chemical or physical property of interest. Some data collected by Bilonick (1989) on percent ash in coal and shown in Figure B.1 illustrate this phenomenon quite well. Higher-density particles have higher percent ash content regardless of the particle volume. On the other hand, higher-volume particles with the same density have about the same percent ash content.

Under these circumstances, and using the fact that mass equals density times volume (size), Gy further simplifies (B.3):

$$\text{(B.4)} \qquad \text{Var(FE)} \approx (1/M_S - 1/M_L)\ (\text{Size Factor})\ (\text{Density Factor}).$$

The size factor is approximated by the product of the third power of the near maximum particle size d,[21] a shape (or form) factor f, and a particle size distribution (granulometric) factor g. The density factor is approximated by the product of a composition factor c and a "liberation" factor l. A summary of common or calculated values for these parameters is given in Table B.1 (Pitard,

[21] The value d is the size of the opening of a screen retaining 5% by weight of the lot to be sampled. The logic behind the selection of d is based on the principle of particle separation using a sieve. Holes in the sieve are square shaped. A cube-shaped particle with each side having length d will exactly fit through a sieve whose squares also have side length d. Particles smaller than a cube of side d, such as spheres with diameter d, or some larger particles, such as cylinders with diameter d, will also fit through this sieve.

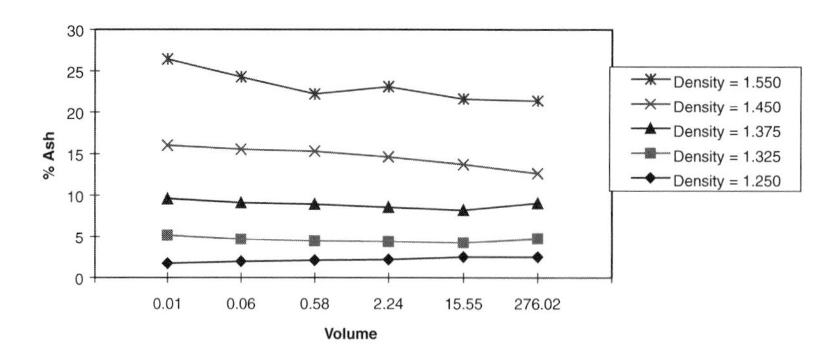

Figure B.1: Percent ash varies more by density than by volume.

1993, pp. 159–163). After substituting for the size and density approximations, (B.4) becomes (Gy, 1992, pp. 86–88)

$$\text{Var(FE)} \approx (1/M_S - 1/M_L)(d^3 fgcl). \tag{B.5}$$

Because of the estimations used, *this formula provides only an order of magnitude approximation* of the relationship between the minimum possible sampling variation, the particle size, and the sample mass. This formula is a first-order approximation and does not apply to trace amounts (Gy, 1992, p. 360). Under appropriate circumstances, however, by characterizing the material using estimates of these factors, this approximation can be used in practice in several ways. First, note that because the maximum particle size d is directly proportional to Var(FE), a reduction in particle size by grinding the entire lot will reduce the Var(FE). (This is discussed in detail later in this appendix.) We can also estimate the minimum sampling-to-sampling variation (even though we cannot repeat the sampling protocol) for the particle size we have and for the sample mass we currently take. We can thus assess whether the current sampling protocol is acceptable. We can also design a sampling protocol by estimating the particle size reduction and sample mass needed to obtain a desired sample precision.

B.5 Approximation for particle size distribution

When sampling to determine particle size distribution, the sampling variation will generally be different for each class size. The formula must therefore be recalculated for each class. For this application, c_L in (B.3) represents the true content fraction of the class of interest. The value c_i of an individual particle is either 1 or 0 depending on whether the ith particle is in this class or not. As in the general case discussed above, size and density approximations are used to

Table B.1: Common values of factors for the FE.

Shape (Form) Factor

f = (Vol. of particle with diameter d)/(Vol. of cube of side d)

Shape	Value	Comment
Flakes	0.1	
Nuggets	0.2	
Spheres	0.5	most common
Cubes	1	basis for calculations
Needles	[1, 10]	Length divided by width. In the formula above, d = diameter of needle

Granulometric Factor (Size Distribution)

g = (diameter of smallest 5% of material)/(diameter of largest 5% of material)

Type	Value	Comment
Noncalibrated	0.25	from a jaw crusher
Calibrated	0.55	between 2 consecutive screen openings
Naturally Calibrated	0.75	cereals, beans, rice, ...
Perfectly Calibrated	1	ideal

Liberation Factor

l = (Maximum content − ave. content)/(1 − ave. content) or l = {(diameter at which particle is completely liberated)/(current diameter)}^0.5

Type	Value	Comment
Almost Homogeneous	0.05	not liberated
Homogeneous	0.1	
Average	0.2	
Heterogeneous	0.4	
Very Heterogeneous	0.8	liberated

Mineralogical (Composition) Factor (g/cm³)

$$c = [(\lambda_m(1-c)^2)/c] + [\lambda_g(1-c)] \cong \begin{cases} \lambda_m/c & \text{for } c < 0.1, \\ \lambda_g(1-c) & \text{for } c > 0.9, \end{cases}$$

where

λ_m = density of the material of interest (in g/cm^3),
λ_g = density of everything but the material of interest (in g/cm^3),
c = content as a weight proportion of the material of interest.

estimate the ith particle mass M_i. Making these substitutions and using λ to represent density, (B.3) becomes (Pitard, 1993, p. 334)

$$\text{Var(FE)} \approx (1/M_S - 1/M_L)f\lambda[(1/c_L - 2)d_1^3 + gd^3], \tag{B.6}$$

where d_1 is the average particle size for the class of interest and d the near maximum particle size for all other classes combined.

Particle size distribution analysis is important for many commercial plants, and (B.6) can be used to calculate the sample mass needed to obtain a specified sampling precision. Also, for a sample to be representative of the lot, it must be representative of the particle size distribution. So sample mass estimations here can be used to test for accuracy.

B.6 Other applications

The approximate relationship between the constitution heterogeneity (CH) and the statistical variance, discussed in the next section, can be used to determine the optimal sample mass when sampling to determine the relative amount of trace constituents in the lot. Two series of samples of large and small sample masses are collected and analyzed. The statistical coefficients of variation are calculated. These, together with the values of the sample mass, can be used to determine the minimum sample mass needed to obtain the desired sampling precision of the estimated critical content of the trace constituent. For details, see Chapter 20 of Pitard (1993).

B.7 The effects of particle size reduction (grinding)

Hypothetical, unrealistic case

If all particles were alike, then the CH would be zero. Since there would be no variation in sampling, the Var(FE) = 0, also. Particle size reduction by grinding might separate out nuggets of material with the property of interest. In that case, the CH and the Var(FE) would no longer be zero. Both would increase. Figure B.2 illustrates this phenomenon.

This is a situation that will rarely, if ever, occur in practice, and the approximating formulas for Var(FE) do not apply here.

Practical case

We can see from the approximating formulas (B.5) and (B.6) that if the maximum particle size d is reduced, then the Var(FE) is also reduced. We illustrate this with a simplified, specific example. In Table B.2 are the critical contents c_i listed in increasing order and the corresponding heterogeneities h_i of 8 particles that constitute an entire lot.

The h_i values are calculated assuming equal mass for every particle, that is $M_i = M_L/N_L$. Thus, from (B.1),

$$h_i = (c_i - c_L)/c_L. \tag{B.7}$$

The statistical mean and variance of the true lot content are 10.00 and 28.06, respectively. The CH = Var(h_i) = 0.28. If we sample $n = 1$ particle from the

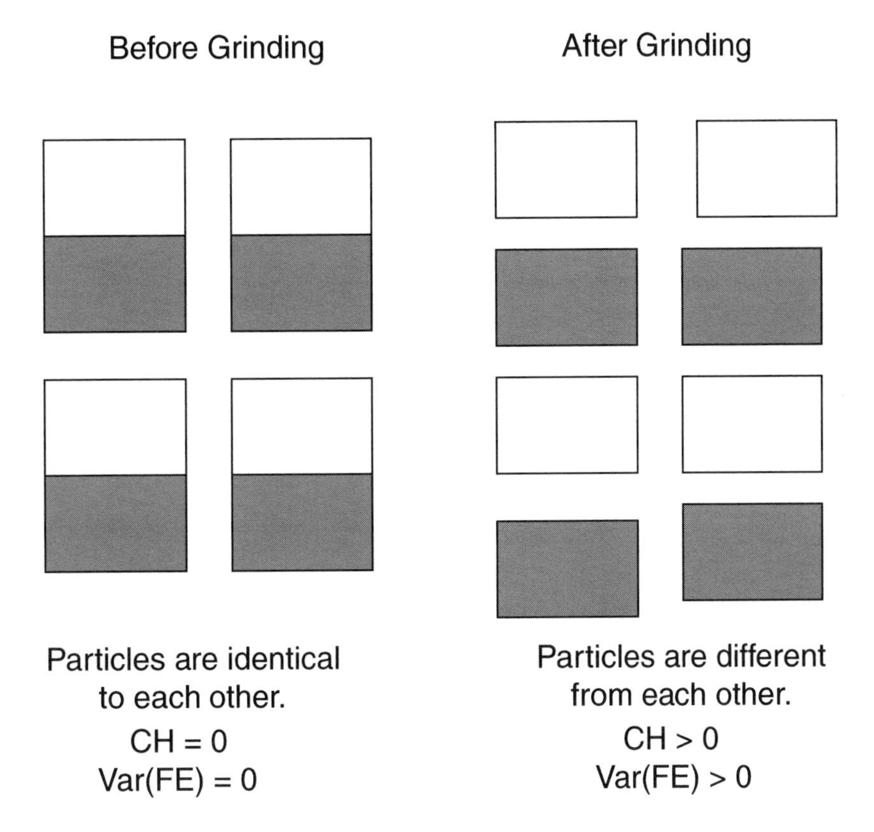

Figure B.2: Both the CH and the Var(FE) increase after grinding in a hypothetical, unrealistic case.

$N_L = 8$ particles in the lot, then the classical statistical probability of selecting any given particle is $P = 1/8$. Thus, from the first part of (B.3),

$$\text{Var(FE)} \approx \left[\frac{1-(1/8)}{(1/8)*8}\right] * 0.28 = 0.24.$$

Now suppose we grind each of the original 8 particles into 6 smaller particles and end up with a total of 48 particles. Their critical contents and heterogeneities are given in Table B.3.

The statistical mean of the true lot content is still 10.00, but the variance is now 47.36.

The CH $= \text{Var}(h_i) = 0.47$, which is more than 2/3 higher than before grinding, from 0.28 to 0.47. So grinding has increased the CH. However, if we select $n = 6$ particles at random to constitute the sample so that we have the same sample mass as before, then the probability that any given particle will be in the sample is $P = 6/48$. Thus, from the first part of (B.3),

$$\text{Var(FE)} \approx \left[\frac{1-(6/48)}{(6/48)*48}\right] * 0.47 = 0.07$$

Table B.2: Statistics on original lot of eight particles (results rounded to two decimal places).

	Critical Contents (c_i)	Heterogeneities (h_i)	
	3.17	−0.68	
	6.17	−0.38	
	7.67	−0.23	
	8.50	−0.15	
	9.67	−0.03	
	11.00	0.10	
	13.00	0.30	
	20.83	1.08	
Mean(Lot)	10.00		
Variance(Lot)	28.06	0.28	CH = Var(h_i)
Number N_L in Lot	8		
Number n in Sample	1		
Probability P any Particle is in the Sample	1/8		
Var(Mean) = Var(Lot)/n	28.06		
Rel.Var(Mean) = Var(Mean)/Mean^2	0.28	0.28	CH/n

So, in this case, grinding reduces the Var(FE) by more than 2/3, from 0.24 to 0.07. Repeating the sampling protocol, the sampling-to-sampling variation is thus smaller.

B.8 Comparison of measures of variation

Let's compare these results to a well-known statistical measure, the variance of the (sample) mean and the relative variance of the (sample) mean, defined as

(B.8)

$$\text{Variance of the sample mean} = \frac{\text{Var (a one-particle sample)}}{\text{\# particles in a several-particle sample}}$$

and

(B.9)

$$\text{Relative variance of the sample mean} = \frac{\frac{\text{Var (a one-particle sample)}}{\text{\# particles in a several-particle sample}}}{\text{mean}^2}.$$

For the original lot of 8 particles, we have from (B.9)

$$\text{Original lot: Statistical rel. var. (sample mean)} = \frac{(28.06/1)}{10^2} = 0.28.$$

Table B.3: Statistics on ground lot of 48 particles (results rounded to 2 decimal places).

	Critical Contents (c_i)						Heterogeneities (h_i)					
	1	2	3	4	4	5	−0.9	−0.8	−0.7	−0.6	−0.6	−0.5
	5	6	6	6	7	7	−0.5	−0.4	−0.4	−0.4	−0.3	−0.3
	7	7	8	8	8	8	−0.3	−0.3	−0.2	−0.2	−0.2	−0.2
	8	8	8	9	9	9	−0.2	−0.2	−0.2	−0.1	−0.1	−0.1
	9	9	10	10	10	10	−0.1	−0.1	0	0	0	0
	10	11	11	11	11	12	0	0.1	0.1	0.1	0.1	0.2
	12	12	13	13	14	14	0.2	0.2	0.3	0.3	0.4	0.4
	14	15	15	15	16	50	0.4	0.5	0.5	0.5	0.6	4
Mean(Lot)	10											
Variance(Lot)	47.36						0.47	CH = VAR(h_i)				
Number N_L in Lot	48											
Number n in Sample	6											
Probability P any Particle is in the Sample	6/48											
Var(Mean) = Var(Lot)/n	7.89											
Rel.Var(Mean) = Var(Mean)/ Mean^2	0.08						0.08	CH/n				

For the same lot with the 8 particles ground to 48 and a sample with 6 particles, we have

$$\text{Ground lot: Statistical rel. var. (sample mean)} = \frac{(47.36/6)}{10^2} = 0.08.$$

The statistical variance of the critical content was higher in the second case, similar to the situation with CH. However, the statistical relative variance of the mean was lower, similar to the situation with the Var(FE).

Substituting (B.7) into (B.2) yields

$$\text{CH} = \text{Var}(h_i) = \text{Var}(c_i - c_L)/c_L^2 = \text{Var}(c_i)/c_L^2 = \text{Statistical rel. var.}$$

Dividing both sides by the number of particles in the sample and using (B.8) and (B.9), we have

$$\text{CH}/n = \text{Statistical rel. var. (sample mean)}.$$

When the probability of selection is constant so that $P = n/N_L$, Stenback (1998) has noted from the first part of (B.3) that

$$\begin{aligned}\text{Var(FE)} \approx [(1-P)\text{CH}]/(P * N_L) &= \frac{(1-n/N_L)\text{CH}}{N_L * n/N_L} \\ &= \frac{N_L - n}{N_L} * \frac{\text{CH}}{n} \\ &= \frac{N_L - n}{N_L} * \text{Statistical rel. var. (sample mean)}.\end{aligned}$$

The quantity $[(N_L - n)/N_L]$ is known as a *finite population correction factor.* When the number N_L of particles in the lot is very large relative to the number n of particles in the sample, $[(N_L - n)/N_L] \approx 1$. The Var(FE) will then be approximately equal to the statistical rel. var. (sample mean).

In other words, when the lot is large relative to the sample and the particle masses are about the same, then the statistical relative variance of the mean and Gy's variance of the FE will be very close numerically.

Appendix C

Obtaining a Sequential Random Sample

In processes where solid material is fed into containers such as bags or drums at the end of the production line, it may not be logistically feasible or it may take too much time to wait until all containers are filled to select a random sample. The containers would have to be stored somewhere until all of them were filled. Then those selected for the sample would have to be retrieved. Or if the process were continuous and we shipped the product as soon as a truck could be loaded, then we would never have an entire lot from which to sample. The following method can be used for obtaining a random sample of discrete units when the lot items are accessible sequentially (Kennedy and Gentle, 1980).

Suppose we have a batch process in which bags are filled with product at the end of the line, and we would like to sample 1% of the bags. Then each bag should have probability 0.01 of being selected. Using a computer, we generate a series of random numbers[22] between 0 and 1, and we generate more numbers than we think there will be bags. Often, we do not know exactly how many bags there will be in a lot, but based on past experience we have a good idea of what the maximum number will be. We can even generate the random numbers on the spot, as would be necessary for a continuous process in which we keep filling and shipping bags and never have a defined lot.

As each bag is filled, we examine the next random number in the sequence and apply the following rule.

1. If the random number is less than or equal to 0.01, the bag is selected as part of the sample and set aside.

2. If the random number is greater than 0.01, the bag is not selected as part of the sample and is shipped or stored with the rest of the lot.

[22]With single-precision computations, numbers with at least seven decimal places will be generated. Many statistical software packages give six places. This should be sufficient for most purposes.

Since approximately 1% of the random numbers will be less than or equal to 0.01 by design, approximately 1% of the bags will be chosen to be in the sample. While this technique will not guarantee a sample of exactly 1%, it will generally come fairly close. If a different percent, say $p\%$, is required for the sample, the cutoff point for the random numbers is $p/100$.

Kennedy and Gentle also give a procedure that guarantees to the nearest integer an *exact percent*. It is more complicated to execute in practice than the previous method because it requires taking an initial sample and continuously switching out bags when certain random numbers come up. Unfortunately, this is not feasible logistically if the number of units in the sample is large.

Appendix D

Calculation of the Variogram

Suppose samples are taken on a fixed schedule a certain number of time units apart: every hour, for example, at the beginning of each shift, or at noon every day. For the ith sample, a property of interest (the critical content) is measured as a percent weight *fraction*[23] *of the total* sample weight. This content, c_i, is recorded. After a fixed number n of samples is taken, we compute the average A of the contents c_i. Now we are ready to compute each point in the variogram. The jth variogram point $V(j)$ is calculated as

$$V(j) = \left[\sum_{i=1}^{n-j} (c_{i+j} - c_i)^2 \right] \Big/ \; [2(n-j)A^2].$$

This is the variation of all samples that are j time units apart. Division by A^2 makes it a relative variance. An example is given in Figure D.1 with code for the Excel macro given in Table D.1.

[23] Fractions are used to allow comparison of the nugget-effect value to the estimated material variation.

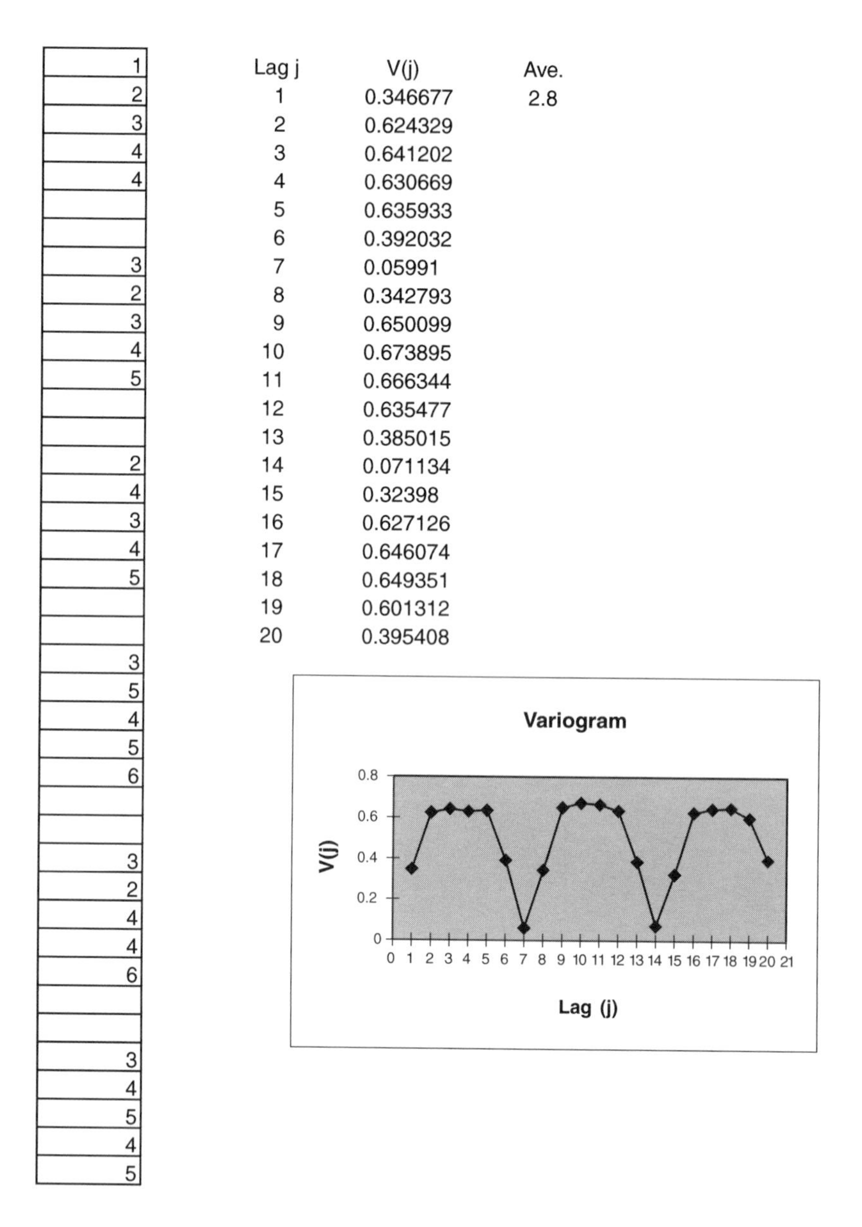

Data
1
2
3
4
4
3
2
3
4
5
2
4
3
4
5
3
5
4
5
6
3
2
4
4
6
3
4
5
4
5

Lag j	V(j)	Ave.
1	0.346677	2.8
2	0.624329	
3	0.641202	
4	0.630669	
5	0.635933	
6	0.392032	
7	0.05991	
8	0.342793	
9	0.650099	
10	0.673895	
11	0.666344	
12	0.635477	
13	0.385015	
14	0.071134	
15	0.32398	
16	0.627126	
17	0.646074	
18	0.649351	
19	0.601312	
20	0.395408	

Figure D.1: Data and variogram for Figure 4.17. Blanks indicate missing values.

Table D.1: Visual basic code for variogram.

```
' Code developed and Copyrighted by Windward Technologies, Inc.
' Licensed to Alpha Stat Consulting Company for individual use only.
' Contact Alpha Stat Consulting Company for other licensing options.
' Web: www.alphastat.com   Email: alpha@alphastat.com

Option Explicit
Option Base 1

Dim xd As Range
Dim yd As Range

Dim y(4) As Integer
Dim s As Range

Sub Variogram()
Dim x(4) As Integer, num As Integer, num2 As Integer
Dim data() As Double, v() As Double, ave As Double, ave2 As Double, varsum As Double
Dim denom As Double
Dim i As Integer, j As Integer, special_const As Integer, num_non_numeric
Dim chart_name As String
Dim msg As String, Style, Title As String, response

special_const = -31415
Set s = Cells(1, 1)

Set xd = Selection
    x(1) = Selection.Row
    x(2) = x(1) + Selection.Rows.Count - 1
    x(3) = Selection.Columns.Count
num = x(2) - x(1) + 1
num2 = num / 2

' We compute num2 lags
' If num - 30 > num/2 then we compute num - 30 lags
' otherwise we compute num/2 lags
If (num - 30 > num2) Then
    num2 = num - 30
End If

If (num < 2) Then
        msg = "There must be at least two rows"
        Style = vbCritical
        Title = "Not enough rows"
        response = MsgBox(msg, Style, Title)

        GoTo 9999
End If

If (x(3) <> 1) Then
        msg = "Selected data must be in exactly one column"
        Style = vbCritical
        Title = "One Column"
        response = MsgBox(msg, Style, Title)

        GoTo 9999
End If
```

Table D.1: (*Continued*)

```
ReDim data(num), v(num2)
ave = 0
num_non_numeric = 0
For i = 1 To num
    If (Not IsNumeric(xd(i))) Then
        data(i) = special_const
        num_non_numeric = num_non_numeric + 1
    Else
        data(i) = xd(i)
        ave = ave + data(i)
    End If

Next
' Error checking
If (num - num_non_numeric = 0) Then
        msg = "There is NO numeric data"
        Style = vbCritical
        Title = "Numeric data necessary"
        response = MsgBox(msg, Style, Title)

        GoTo 9999
End If



ave = ave / (num - num_non_numeric)

If (Abs(ave) <> 0#) Then
       ave2 = ave * ave
Else
     msg = "A variogram cannot be computed, since the mean is zero."
        Style = vbCritical
        Title = "Mean Zero"
        response = MsgBox(msg, Style, Title)

        GoTo 9999
End If

For j = 1 To num2
    varsum = 0
    denom = num - j
    For i = 1 To num - j
        If (data(i + j) = special_const Or data(i) = special_const) Then
        denom = denom - 1
        Else
        varsum = varsum + (data(i + j) - data(i)) ^ 2
        End If
    Next
    If (denom = 0) Then
        v(j) = -1
    Else
        varsum = varsum / (2 * ave2 * denom)
        v(j) = varsum
    End If
```

Table D.1: (*Continued*)

```
Next

Application.ScreenUpdating = False

For i = 1 To num2
     s.Cells(i + 1, 8) = i
    s.Cells(i + 1, 9) = v(i)
Next
    s.Cells(2, 10) = ave

s.Cells(1, 8) = "Lag j"
s.Cells(1, 9) = "V(j)"
s.Cells(1, 10) = "Ave."

      ActiveSheet.ChartObjects.Add(67.2, 114, 250.8, 186.6).Select
    Application.CutCopyMode = False
    ActiveChart.ChartWizard Source:=Range(Cells(2, 8), Cells(num2 + 1, 9)), _
        Gallery:=xlXYScatter _
        , Format:=2, PlotBy:=xlColumns, CategoryLabels:=1, _
        SeriesLabels:=0, HasLegend:=2, Title:="Variogram", _
        CategoryTitle:="Lag (j)", ValueTitle:="V(j)", ExtraTitle:=""

    i = ActiveSheet.DrawingObjects.Count
    chart_name = ActiveSheet.DrawingObjects(i).Name
    ActiveSheet.ChartObjects(chart_name).Activate
     'ActiveSheet.ChartObjects("Chart 48").Activate
    ActiveChart.Axes(xlCategory).Select
    With ActiveChart.Axes(xlCategory)
        .MinimumScale = 0
        .MaximumScaleIsAuto = True
        .MinorUnit = 0.2
        .MajorUnit = 1
        .Crosses = xlAutomatic
        .ReversePlotOrder = False
        .ScaleType = False
 End With

 Application.ScreenUpdating = True

 9999
 End Sub
```

Appendix E

Experiments

Below are three experiments that you can perform easily, even at home. They illustrate some of the issues discussed in Chapter 3. The first addresses sampling dimension, the second solids sampling, and the third liquid sampling.

Experiment 1: Sampling Dimension

Get 27 building blocks or facsimile. Vary the colors or design of the blocks as much as possible. Put them in stacks of 3 and arrange the stacks to form a $3 \times 3 \times 3$ cube, as in Figure E.1.

The analogy is a three-dimensional pile of coal, catalyst, plastic pellets, soil particles, or other solid material. It could be in the field, plant, or lab. From each of these perspectives, consider how you would physically obtain several randomly chosen individual blocks from this group of 27. How would you access material in the middle, realizing that the material readily accessible on the top or outside may not be representative? How would you define a correct sample and extract it? Next, reduce the sampling dimension to 2 or 1. Define and extract correct samples. How well would this work in situations you encounter? Review your methods and logic after segregating the blocks by color from top to bottom or using some other pattern.

Experiment 2: Solids

Get two or three different types of dried beans. Lentils, black beans, and garbanzo beans work well because they have different sizes, shapes, and densities. Weigh the amount of each bean type. Compute and record the percent weight of each bean type for the combination.[24] Consider these the true percent weights. You will compare results from various samples to the true values to see how representative your samples are.

[24]For example, if you have 50 g of lentils, 250 g of black beans, and 200 g of garbanzo beans, then the total weight is $50 + 250 + 200 = 500$ g. The percent weights are $50/500 = 10\%$, $250/500 = 50\%$, and $200/500 = 40\%$, respectively.

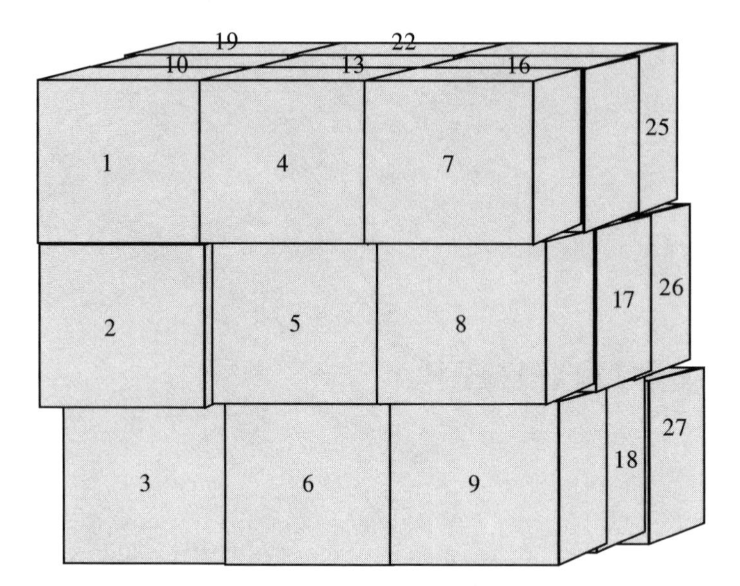

Figure E.1: 27 numbered blocks arranged in a cube.

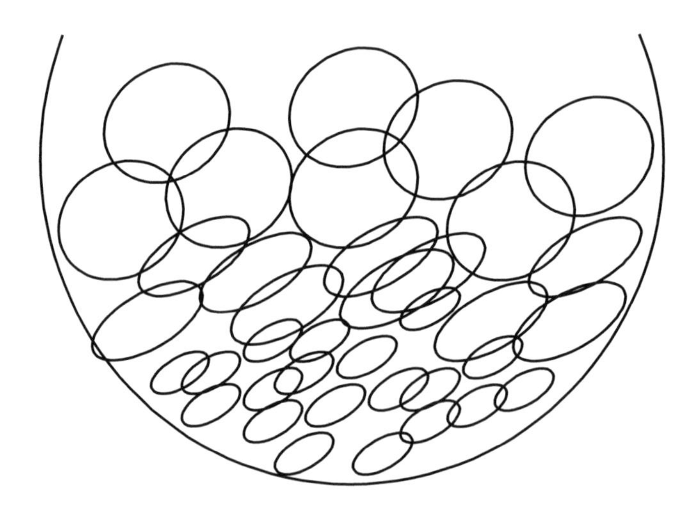

Figure E.2: Bowl of three kinds of beans.

Combine all the beans in a bowl, plastic bag, or other container to form the lot from which you will sample, as shown in Figure E.2.

Mix the beans by stirring or shaking. You will probably have only limited success since smaller, lighter-weight beans will move to the bottom. Decide on a convenient sample weight, and use it for all samples. It's all right if the weights vary a little from sample to sample.

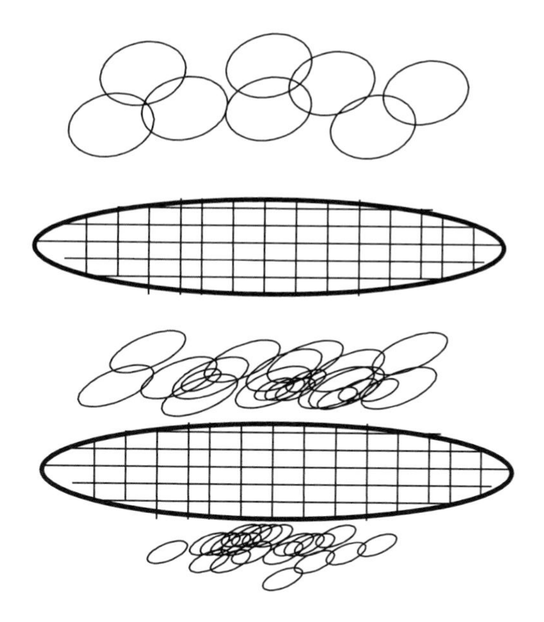

Figure E.3: Separating the three types of beans.

Repeat the sampling protocol described below several times, taking samples using different, but not necessarily correct, sampling techniques, such as pouring, scooping, and slicing across a flattened pile. Use a riffler for one of your sampling methods if one is available. You can also use other techniques you are familiar with or those used in your lab. Take at least three samples using each technique, returning each sample to the original mixture before taking the next sample. With scooping and slicing you will probably need to take several increments and composite them to get the desired sample weight. Of course, we know from Chapter 2 that taking several increments and compositing them is better anyway. Note how difficult it is in some cases to define a correct sample and how easy it is to make an extraction error. Also, observe with scooping and slicing how segregation increases as you take the increments.

After taking each sample, separate the types of beans in the sample. You can do this by hand or by using a sieve or some type of screening device, as shown in Figure E.3. A cooking grill or screen for cooling pizza will work for some sizes of beans. If you use a screening device, note how broken beans of one size can get mixed up with beans of another size. This is a handling error since the weight of a part of the sample of interest has changed through addition or loss.

After separating the beans in the sample, determine and record the percent weight of each bean type using a table such as that shown in Figure E.4. You will have a table like this for each type of sampling technique.

SAMPLE	1		2		3		Summary	
Bean Type	Weight	% Wt.	Weight	% Wt.	Weight	% Wt.	Ave. %	SD %
Garbanzo								
Black								
Lentils								
TOTAL								

Figure E.4: Statistics on the three samples.

You can use a laboratory scale if available to determine the weights, but a food scale will serve our purposes.[25] Compare weight percents for the three samples. Then, for each bean type, compute a mean and standard deviation (SD) of percent weights for these samples.[26] After you have repeated the sampling protocol for different sampling techniques, observe the differences in precision and accuracy between different sampling methods. The more correct the sampling tool and technique, the closer the sample results should be to the true value. A riffler should produce more representative samples than scooping or pouring, for example. Pouring will probably be the least accurate.

To extend the experiment further, take a subsample from one of the samples. This is called *two-stage sampling*. An illustration is shown in Figure E.5. Chemical and physical analyses generally require very small amounts of the sampled material, so subsampling in the lab is routine. This is one of the reasons chemists are good resources for sampling techniques. Now, even if your second-stage sample is representative, it is only representative of the *first-stage sample*, not the *lot*. Compare the percent weight from the second-stage sample of each bean type to the first-stage sample from which it was taken and also to the entire lot. Repeat this several times.

You can vary this experiment in other ways. Try taking just one scoop or slice for the sample rather than several increments, and note the variation between the samples taken using the same technique. You can also take larger or smaller weight samples. If the tools and techniques follow the principle of correct sampling, larger weight samples should have less variation since we saw in Chapter 2 that the variance of the FE will be smaller. You can also put the entire mixture in a closed container. This simulates sampling from drums, silos, or processing units where you can't see the lot and blissfully sample from the top, side, or bottom.

Experiment 3: Liquids

Take a small bottle of mineral oil and pour out some of it. It doesn't really matter how much you pour out. Fill the bottle back to volume with water. The water and oil do not mix. Add a couple of drops of food coloring, something

[25] Using a four-place balance will give very accurate weights for the sample. However, if the sample isn't representative of the lot, it doesn't help much and gives a false sense of security.

[26] The average percents and SDs will not be statistically "best," but if the sample weights are roughly the same, they will be good enough for illustration purposes here.

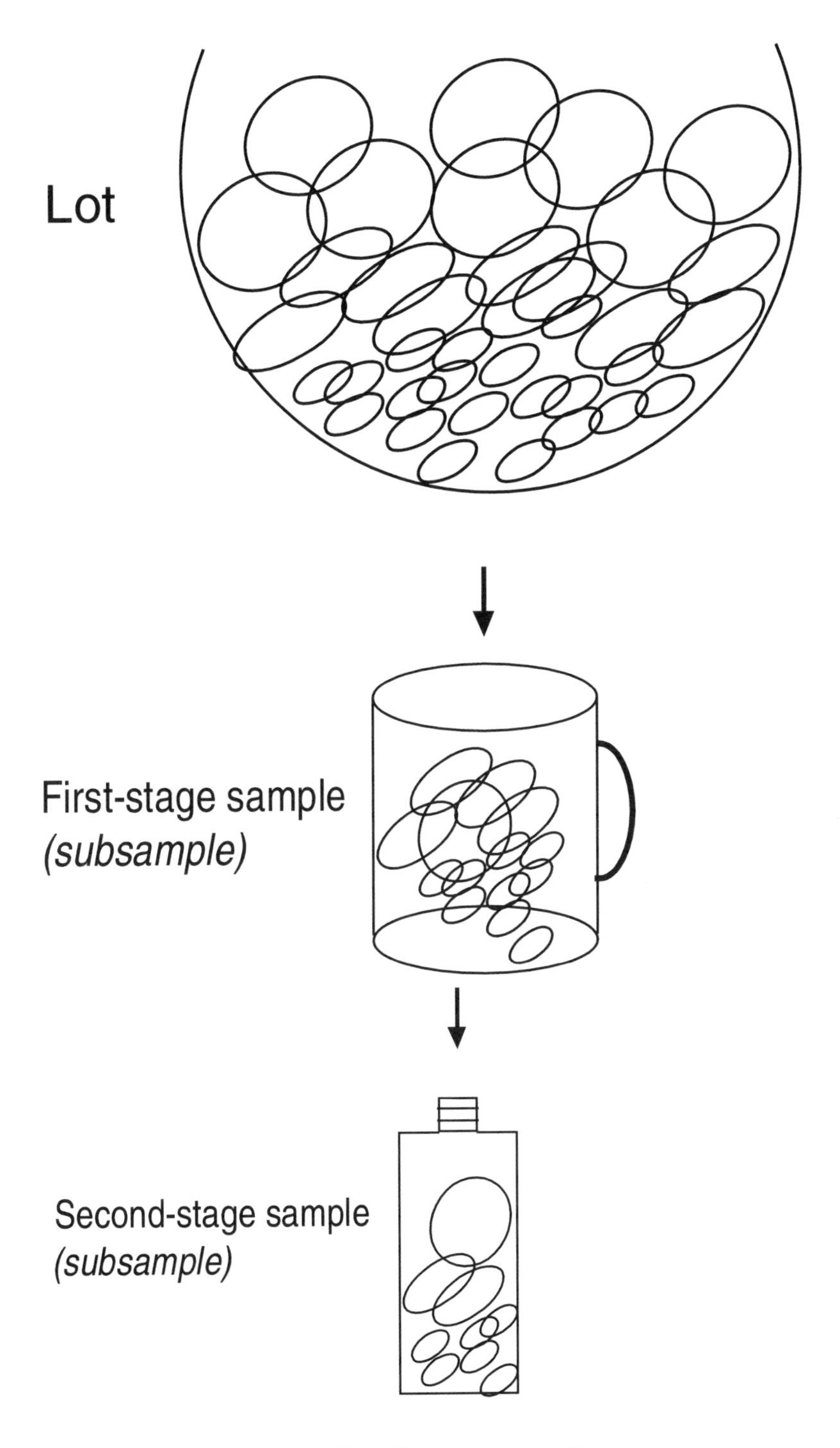

Figure E.5: Two-stage sampling.

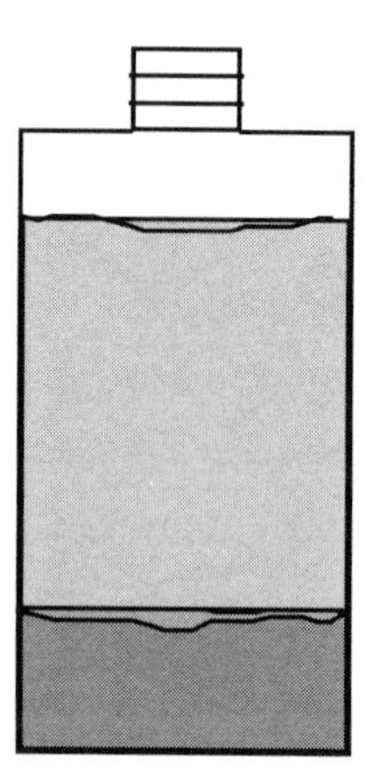

Figure E.6: Water and oil do not mix.

dark like red, green, or blue. The food coloring will mix with the water but not the oil, so the two liquids will be easy to identify.

Shaking the bottle to mix the water and oil does not work very well, and most of the water settles out within a few seconds, as shown in Figure E.6. Within a few minutes, the two liquids are almost completely separated again. How would you get a representative sample from this mixture? Clearly, pouring from the top will give mostly oil, even after shaking. Using a pipette will only get one liquid component. A straw lowered slowly to the bottom will fill with both liquids, and capping it with your finger and pulling it up might work. With low-viscosity liquids, however, the liquid may drip out of the straw immediately before you are able to transfer it to another container. This scenario simulates a lab sample and the difficulty the lab may have in subsampling to get an amount suitable for a chemical analysis.

Now generalize this scenario, and consider how to sample a large liquid-filled tank or rail car, where shaking to mix is not possible. If the liquids were mixed using a mechanical device, how could you evaluate its effectiveness? If you took a sample from a spigot near the bottom, how good would the sample be? How would you sample if the liquids were flowing in a pipe? A review of Chapter 3 may help you answer these questions.

Appendix F

Sampling Safely

F.1 Introduction

When doing any kind of sampling, you must provide for your personal safety. You should know the dangers and hazards of the *material* you will sample, the *equipment* you will use, and the *setting* where you will take the sample. In industrial settings, government regulations require minimal safety training and protective equipment. Personal protective clothing and equipment should be worn not to satisfy government regulations, however, but to protect yourself. Make sure you get instructions. Whether you are a company insider or outsider, you should not perform the sampling if you have doubts about your personal safety. You are best served by knowing what to look for and what questions to ask, by asking those questions, and then by making an informed decision.

This discussion is not designed to provide, nor can it provide, all the safety information you may need in various circumstances. There are too many specialized situations. The purpose here is to raise awareness and give some basic information.

F.2 The material

Sampling extremely hot or cold material requires gloves for hand protection. Goggles for eye protection should be worn when sampling liquids that may splash or solids that may generate dust. A nose and mouth cover, like those worn by surgeons, can protect you from ingesting dust. A full face mask will be necessary near poisonous gases in a chemical or biological environment.

F.3 The equipment

Special containers should be used for collecting samples under pressure. Ensure that the equipment and required connections are working properly. Get the proper instructions and training before proceeding. Glass or plastic vials should not be used for collecting material at extremely high or low temperatures since they may shatter or melt.

F.4 The setting

Industrial and mining settings should require ear protection (plugs or muffs) for noise above a certain decibel level, hard hats for head protection, and steel-toed shoes for foot protection. A safety harness should be worn when sampling from heights. A full protective suit should be worn when in extreme heat or cold or near hazardous chemicals. Check to see whether mechanical equipment must be shut down to sample safely. Eye or face protection should be worn if hot gases may be released to the atmosphere. In some circumstances, weather may be a factor. In the field, local terrain may require special gear.

References

Allen, T. and Khan, A. A. (1970). "Critical Evaluation of Powder Sampling Procedures," *Chemical Engineer*, Vol. 238, p. 108–112.

ASTM D 4057 (1981). "Standard Practice for Manual Sampling of Petroleum and Petroleum Products." ASTM Committee on Standards, 1916 Race St., Philadelphia, PA 19103.

ASTM D 4177 (1982). "Standard Method for Automatic Sampling of Petroleum and Petroleum Products." ASTM Committee on Standards, 1916 Race St., Philadelphia, PA 19103.

Bilonick, Richard A. (1989). "Quantifying Sampling Precision for Coal Ash Using Gy's Discrete Model of the Fundamental Error," *Journal of Coal Quality*, Vol. 8, pp. 33–39.

Cochran, William G. (1977). *Sampling Techniques*, Third Edition, New York: Wiley.

Cressie, Noel (1993). *Statistics for Spatial Data*, New York: Wiley.

Duncan, Acheson J. (1962). "Bulk Sampling: Problems and Lines of Attack," *Technometrics*, Vol. 4, pp. 319–344.

Elder, R. S., Thompson, W. O., and Myers, R. H. (1980). "Properties of Composite Sampling Procedures," *Technometrics*, Vol. 22, pp. 179–186.

Francis Pitard Sampling Consultants, LLC. (1995). Software: *Process Variability Management.* Bloomfield, CO.

Gy, Pierre M. (1992). *Sampling of Heterogeneous and Dynamic Material Systems: Theories of Heterogeneity, Sampling and Homogenizing*, Amsterdam: Elsevier.

Gy, Pierre M. (1998). *Sampling for Analytical Purposes*, Chichester: Wiley.

Johanson, Jerry R. (1978). "Particle Segregation and What to Do About It," *Chemical Engineering*, May 8, pp. 183–188.

Kennedy, William J., and Gentle, James E. (1980). *Statistical Computing*, New York: Marcel Dekker.

Leutwyler, Kristin (1993). "Shaking Conventional Wisdom," *Scientific American*, September, p. 24.

Merriam-Webster's Collegiate Dictionary, Tenth Edition (2000). Springfield, MA: Merriam-Webster, Inc.

Pitard, Francis F. (1993). *Pierre Gy's Sampling Theory and Sampling Practice: Heterogeneity, Sampling Correctness, and Statistical Process Control*, Second Edition, Boca Raton: CRC Press.

Smith, R., and James, G. V. (1981). *The Sampling of Bulk Materials*, London: The Royal Society of Chemistry.

Stenback, G. A. (1998). Personal communication, Iowa State University, Ames, IA.

Studt, Tim (1995). "For Material Researchers, It's Back to the Sandbox," *R&D Magazine*, July, pp. 41–42.

Wallace, Dean, and Kratochvil, Byron (1985). "Use of a Mill and Spinning Riffle for Subsampling Laboratory Samples of Oil Sand," *Aostra Journal of Research*, Vol. 2, pp. 233–239.

Welker, T. F. (1984). "Crude Oil Sampling for Custody Transfer," contained in advertising information for Welker Engineering Company, Sugar Land, TX.

Welker, T. F. (1989). "Definitions of Heating Value," Internal memo to sales and service representatives of Welker Engineering Company, June 7, Sugar Land, TX.

Welker, T. F. (1997). "Wet Gas Sampling," contained in advertising information for Welker Engineering Company, Sugar Land, TX.

Index